Study Guide and Computer Workbook
to accompany

LOCKHART'S

INTRODUCTION TO STATISTICS AND DATA ANALYSIS IN THE BEHAVIORAL SCIENCES

Robert Lockhart

Philip Groff

W. H. Freeman and Company
New York

ISBN 0-7167-3197-5

Printed in the United States of America

W. H. Freeman and Company
41 Madison Avenue, New York, New York 10010
Houndmills, Basingstoke RG21 6XS, England

First printing 1997

CONTENTS

Preface

Part I Study Guide and Exercises

Chapter 1   Purpose of Statistical Data Analysis

Chapter 2   Graphical and Numerical Descriptions of Data

Chapter 3   Modeling Data and the Estimation of Parameters

Chapter 4   Probability Distributions

Chapter 5   Sampling Distributions and Interval Estimation

Chapter 6   Experiments with Two Independent Groups

Chapter 7   Larger Experiments with Independent Groups:
             Analysis of Variance

Chapter 8   Increasing the Precision of an Experiment

Chapter 9   Quantitative Predictor Variables: Linear Regression
             and Correlation

Chapter 10   Categorical Response Variables and Distribution-Free Methods

Part II Using SPSS and SAS

Introduction to the Use of SPSS for Windows

Section 1   Introduction to SPSS: Reading in Data from Disk

Section 2   Examining and Describing Data

Section 3   Obtaining Standard Errors and Confidence Intervals

Section 4   Analysis of Variance for Independent Groups

Section 5   Matched Pairs and Within-Subjects Designs

Section 6   Regression and Correlation

Section 7   Analysis of Contingency Tables

Introduction to the Use of SAS

Section 1   Introduction to SAS: Reading in Data from Disk

Section 2   Examining and Describing Data

Section 3   Obtaining Standard Errors and Confidence Intervals

Section 4   Analysis of Variance for Independent Groups

Section 5   Matched Pairs and Within-Subjects Designs

Section 6   Regression and Correlation

Section 7   Analysis of Contingency Tables

Appendixes

A Answers to Problems and Exercises

B Data Sets Used in Textbook and Study Guide

PREFACE

This workbook is intended as both a study guide and a source of additional problems and exercises.

Part I takes each chapter of the text and provides

- Brief overview of the chapter

- Point-by-point list of what you should know and be able to do

- Set of short answer questions of the type instructors often use in tests and quizzes

- Set of short problems covering basic concepts

- Exercises in the analysis of realistic data

Part II provides an introduction to the use of two of the most commonly used computer packages: SPSS and SAS. These are offered as independent modules so that if you plan to perform computer analyses you will be able to choose one package and follow a self-contained set of instructions for that package.

In both the SPSS and SAS modules the examples and exercises are based on the data sets used in the text itself. This strategy has the obvious advantages of economy, in that details of exposition of the data sets, and the results of the analyses, are already available. It also has the advantage of linking the computer analyses to the broader discussion of the interpretation of the results.

The data sets for all major examples, problems, and exercises in the text are available to students and instructors in ASCII format. Note that in Part II, the first sections of both SAS and SPSS contain explicit instructions on importing the data files into the respective packages.

Part I of the workbook can be used quite independently of a computer package and therefore of Part II. The issue becomes relevant only for the exercises in the analysis of realistic data. For these exercises, students who are using packages will be expected to import the relevant ASCII data file and complete the appropriate analyses. Students not using packages can make use of the "results of basic computations" that accompany each problem. Such basic computations are typically the means and sums of squares—the heavy arithmetic—which, after the basic concepts have been understood, serves no instructional purpose. This information is also useful to students who are using packages as a means of checking their output.

1 PURPOSE OF STATISTICAL DATA ANALYSIS

Overview

Chapter 1 introduces the elementary concepts that will serve as the foundation of the methods developed in subsequent chapters.

1. Accounting for variability. A major goal of science is to understand variability. How and why do things change? How and why do people differ from each other? Why does the same person display different behavior from one circumstance to another? Do the various conditions of an experiment produce differences among the observed responses? To understand change is to be able to specify the conditions under which it occurs; once such conditions have been specified, future outcomes can be predicted. A statement that makes such a specification is called a *model*.

2. Models and residuals. In the world of real science, models make imperfect predictions. These inaccuracies are reflected in residuals—the differences between a model's prediction and an actual observation. For this reason, residuals are sometimes referred to as errors; they can also be thought of as noise. The larger the residuals the poorer the model's prediction and the greater is the unaccounted for variability. One of the goals of data analysis is therefore to find a model that results in the smallest possible residuals.

3. The general and the particular. The goal of model formulation and data analysis is not only to account for the variability in the data actually obtained. Such data constitute a sample from a population of potential observations. The larger goal is to formulate a model that applies to this population, not just the sample. Included in this idea of the model's generality is the claim that it should apply to repetitions of the experiment over different occasions; the model should be replicable. In practice such repetitions will not yield numerically identical results (strict replication) but are expected to support a model of the same form (general replication).

4. Models and causal mechanisms. A model identifies and describes the regularities in data, the signal imbedded in noise. The purpose of identifying such regularities is not only that of descriptive elegance. Science assumes that regularities are the consequence of causal mechanisms, and that the form of the regularity is an important clue to identifying those mechanisms.

5. Parsimony. Models should be as simple as the data permit. This guiding principle of parsimony is not a claim that nature is simple, but that explanations should not be made more complex than the data demand. For this reason if two different models account for the same amount of variability, then the simpler of the two models will be preferred.

6. Variable types. The most fundamental difference is between response variables and predictor variables. The response variable (denoted Y) is the variable whose values the model attempts to predict. It is often referred to as the dependent variable. The predictor variable (denoted X) is the variable whose values form the basis for predicting values of the response variable. It is the predictor variable that accounts for the variability in the response measure, and it is therefore sometimes referred to as the explanatory variable. It is also called the independent variable. A model can therefore be described as a statement that predicts values of the response variable as a function of values of the predictor variable.

A second distinction is made between categorical and quantitative variables. The values of a categorical variable are labels that distinguish the categories. For example, in an investigation using a categorical predictor variable, the values of the variable are labels designating the conditions or groups that are being compared. The values of a quantitative variable are numbers signifying a quantity. Both predictor and response variables may be either categorical or quantitative.

A third distinction is made between manipulated and natural predictor variable. The behavioral sciences, in common with all sciences, adopt either or both of two strategies for investigating variability. One is to create variability intentionally by manipulating the conditions under which behavior is observed, such as when a participant is given a pre-established quantity of a drug, or placed in one of several training or therapy conditions. In this case the predictor variable is described as manipulated. The second strategy is to observe behavior as a function of naturally occurring variability such as age, IQ, or sex. Notice that both manipulated and natural predictor variables may be either categorical or quantitative.

7. Bias and precision. The goal in designing an experiment is to eliminate bias and to ensure adequate precision. Bias refers to unwanted influences on observations that result in a systematic tendency for observations to be larger (or smaller) than their true value. Precision on the other hand refers to the stability of the observations, to the absence of noise in the data.

8. Randomization and matching. Randomization and matching offer two ways of handling unwanted influences on observations. Randomization eliminates bias by distributing the influences unsystematically across the conditions. Matching, on the other hand, controls unwanted influences by holding them constant, thereby improving precision.

What You Should Know and Be Able to Do

■ Explain the following statements through the use of a simple example:

A model is a statement that specifies the value of a response variable as a function of the values of a predictor variable.

The purpose of a model is to "account for" or "explain" variability.

The more variability that a model accounts for, the more accurate will be its predictions.

Residuals reflect the inaccuracy of a model's predictions.

Other things being equal, the more parsimonious a model the better. Thus scientists tend to adopt the most parsimonious model the data will permit.

■ Given an appropriate description, identify a variable as being:

either the response or predictor variable

either quantitative or categorical

either natural or manipulated.

■ Explain the difference between studies using natural predictor variables and experiments using manipulated predictor variables with respect to the causal interpretation of the results.

■ Describe and distinguish bias and precision as goals of experimental design and explain the relevance of randomization and matching to these goals.

Short-Answer Questions

1. Write a paragraph explaining what it means for a model to "account for variability" using a simple example of your own creation.

2. John is a 24-year-old man who weighs 180 pounds and is 70 inches tall. You are asked to predict John's height and weight in 16 years time, when he will be 40 years old. In which two of these predictions would you have greater confidence? Explain why.

3. Two high school math classes, A and B, that have different teachers but follow the same curriculum, were given a common final exam. The class averages were 74% and 71% for A and B respectively. The following explanations of this difference are proposed:

   a. Class A had brighter students than class B.

   b. Class A had brighter students and a better teacher than class B.

   c. There is no real difference between the two classes. The difference is just a chance event; if the test were given over again there is a 50% chance that Class B would be higher than Class A.

   Rank order these three explanations in terms of parsimony, from least to most parsimonious, and give reasons for your ranking.

4. You want to predict a person's annual income. Think of (a) a categorical and (b) a quantitative predictor variable, the values of which would help you make this prediction. Explain in your own words how these variables would improve your predictions.

5. You are given a choice between two tasks both of which require you to estimate a child's age. You win a prize if your estimate is within 6 months of the correct age. For the first option, the child is chosen randomly from the entire student body of

the local grade school. For the second option, the child is chosen randomly from the Grade 2 class. Which option offers the best opportunity of winning the prize. Explain why.

6. Chapter 1 used the analogy of target shooting to illustrate the concepts of precision and bias. Consider using an automobile's speedometer as a second analogy. Explain the distinction between a speedometer that was biased and one that lacked precision.

Short Problems

Problem 1 In a study designed to examine whether income is indicative of voting preference, the annual income is obtained for each member of a sample of 200 voters. Voters are then asked which of four candidates they plan to vote for in a forthcoming election. Identify the predictor and response variable in this study then decide whether each is categorical or quantitative. Is the predictor variable natural or manipulated?

Problem 2 In a study investigating whether esthetic preference is influenced by educational achievement, the GPA of 100 students is obtained. Students are asked to designate which of five paintings she or he considers the most beautiful. Identify the predictor and response variable in this study then decide whether each is categorical or quantitative. Is the predictor variable natural or manipulated?

Problem 3 To answer this problem read the description of the following three studies.
Study 1 compared the increase in blood pressure of men with that of women upon being exposed to a frightening film scene.
Study 2 investigated whether speed of decision making changes with age by recording the age of 100 volunteers and the time each took to decide between two competing options.
Study 3 compared course marks (expressed as a percentage) in an elementary calculus course of three different groups each using a different study method.

The following alternatives (a) through (h) refer to elements of these studies. Use these alternatives to answer the subsequent questions (i) through (vi).

a. Sex of participant (male versus female, Study 1)

b. Increase in blood pressure (Study 1)

c. A frightening film scene (Study 1)

d. Age (Study 2)

e. Decision time (Study 2)

f. Elementary calculus course (Study 3)

g. Three different study methods (Study 3)

h. Course grades (Study 3)

List any (if there are any) of the above eight alternatives, a through h, that are:

i. Not variables _____

   ii.   Quantitative response variables _____

   iii.   Categorical predictor variables _____

   iv.   Categorical response variables _____

   v.   Manipulated predictor variables _____

   vi.   Natural predictor variables _____

Problem 4 To answer this question read the description of the following three studies.

Study 1 compared the time that one-month old infants looked at their mother's face with the time they looked at their father's face.

Study 2 asked whether height influences earning power by recording the height of 100 men and their annual income for the year 1997.

Study 3 compared effect of three different drugs and a placebo control on simple reaction time.

The following alternatives (a) through (h) refer to elements of these studies. Use these alternatives to answer the subsequent questions (i) through (vii).

   a.   Sex of parent (father versus mother, Study 1)

   b.   Looking time (Study 1)

   c.   One month of age (Study 1)

   d.   Height (Study 2)

   e.   Annual Income (Study 2)

   f.   The year 1997 (Study 2)

   g.   Three drugs and a placebo (Study 3)

   h.   Simple reaction time (Study 3)

List any (if there are any) of the above eight alternatives, a through h, that are:

   i.   Not variables _____

   ii.   Quantitative response variables _____

   iii.   Categorical predictor variables _____

   iv.   Categorical response variables _____

   v.   Manipulated predictor variables _____

   vi.   Natural predictor variables _____

Exercises in the Analysis of Realistic Data

These exercises begin the analysis of a number of data sets that present hypothetical but realistic data from experiments that address questions of substantive interest. As with most of the exercises at the end of each chapter of the textbook, the conclusions to be drawn from these data sets are consistent with published results. Examination of the actual data from such investigations will begin with Chapter 2. In this section we will take a few of these studies and use them to review the basic concepts introduced in Chapter 1

Set 1: Understanding memory

This first set of exercises presents data from experiments designed to examine some of the factors underlying successful remembering. Few skills are more valuable than a good memory, especially to students studying for exams. But what are effective strategies for improving memory? Experiments aimed at understanding the basic processes underlying memory provide many of the answers. Several exercises in the text described experiments that evaluated mnemonic devices for improving memory such as the peg-word method, the use of imagery, and the method of loci. The exercises under the heading "understanding memory" describe experiments that investigate further the factors that influence the effectiveness of remembering.

Exercise 1. Generation effect

One way of trying to remember material is to read it over and over. Is there a better way? The following experiment suggests one such way.

Fifty participants were randomly assigned to one of two conditions. Participants in the *read* condition saw a list of 40 common words such as **small** and **happy** and were instructed to read them out loud in preparation for a subsequent recall test. Participants in the *generate* condition saw a list of words along with the first letter of a second word. They were instructed to generate a word that started with this letter and meant the opposite of the first word. For example, if they saw **large s___**, they responded "small" or if they saw **sad h____** they responded "happy." In this way both groups spoke an identical set of words, participants in the read condition by reading them directly, participants in the generate group by generating the target word in response to the cue word. Following the presentation of the words, all participants were asked to recall as many of the 40 spoken words as they could remember.

a. Identify the predictor and response variable in this study, then decide whether each of these two variables is categorical or quantitative and whether the predictor variable is natural or manipulated.

b. Participants were randomly assigned to one of the two conditions. Suppose instead the investigator had assigned the first 25 volunteers to the generate condition and the next 25 to the read condition. Does this method of assignment (compared to random assignment) leave the experimental design vulnerable to the criticism of bias or a reduction in precision?

c. The number of words recalled for each of the 25 participants in the read condition is as follows.

```
10   7   9  12  10  11  12·  8  14   8  13   6   8
13  11   9  12   9   7  10   7   7  10   5  10
```

A critic describes these data as "noisy," in that they violate strict replication. Explain what the critic means by this comment. What are the most likely sources of noise for such data?

d. Do you judge the following set of data more noisy than the set given in (c), or less?

12 10 10 11 10   8 10   8 12 10   9 10   9
10   8 11 10   9 10 11 10   8   9 12   7

Exercise 2. Memory for chess positions

This study compared the memory of chess experts for chess positions with the memory of novices for the same positions. Two groups were selected. Group A consisted of 22 chess experts, Group B of 22 chess novices. The investigator recorded the number of board positions correctly recalled.

a. Identify the predictor and response variable in this study, then decide whether each of these two variables is categorical or quantitative and whether the predictor variable is natural or manipulated.

b. Suppose that the results from this investigation show that chess experts remember more positions than do novices. The investigator concludes that this superiority is the result of the greater amount of experience that experts have had playing chess. Can you think of an alternative explanation? Why does the study not rule out such an alternative explanation?

Set 2: Child Behavior

The following studies examine various aspects of child behavior, especially those involving parent-child interactions.

Exercise 3. Familiarity and compliance

Are children more likely to comply promptly to a request made by a parent or to one made by a stranger? In this investigation seven-year-old children are taken to a laboratory playroom that has intentionally been left messy. After entering the room the child is asked to clean it up. The time for the child to begin the task is measured. In one condition the child's parent makes the request. In a second condition the request is made by a stranger (the investigator). A total of 56 seven-year-old children participated in the experiment. They were assigned randomly to the conditions, 28 in each.

a. Identify the predictor and response variable in this study, then decide whether each of these two variables is categorical or quantitative and whether the predictor variable is natural or manipulated.

b. Suppose that the results from this investigation show that the children instructed by a stranger start the task more quickly than if asked by the parent. Which of the two following explanations is the more plausible?

  i. Children in the stranger-instructed condition have more compliant personalities than those in the parent-instructed condition.

  ii. Children perceive strangers as more authoritative than parents.

Exercise 4. Authoritarianism

In order to investigate the possible impact if the authoritarianism of mothers on the academic achievement of their children, a researcher selects a sample of 12-year-old children and their academic performance is measured in terms of their average course grade expressed as a percentage. An authoritarianism score is obtained for each child's mother using a multi-item questionnaire.

a. Identify the predictor and response variable in this study, then decide whether each of these two variables is categorical or quantitative and whether the predictor variable is natural or manipulated.

b. Suppose that the results from this investigation show that the children of authoritarian mothers tend to show lower academic performance. The investigator concludes that this relationship shows that authoritarian parents discourage creative thinking in their children. Why might you be suspicious of this conclusion?

Exercise 5. Mood and compliance

Are children more likely to comply promptly to a request if they are in a good mood? In this experiment, as in Exercise 3, seven-year-old children were taken to a laboratory playroom that has intentionally been left messy. In one condition the children were given a small gift and told that they could take it home with them. The child was then asked to clean up the room. Children in a control group received no gift but were simply asked to clean up the room. The time for the child to begin the task is measured.

a. Identify the predictor and response variable in this study, then decide whether each of these two variables is categorical or quantitative and whether the predictor variable is natural or manipulated.

b. The investigator would like to ensure that the only systematic difference between children in the two conditions is their mood. What steps might the investigator take to achieve this goal?

2 GRAPHICAL AND NUMERICAL DESCRIPTIONS OF DATA

Overview

Chapter 2 has three broad goals.

1. Displaying data. A well-trained scientist will examine data in its original form before performing the kinds of analyses that will be described in subsequent chapters. The purpose of this examination is to ensure that the data do not have undesirable features that could call into question the appropriateness of these analyses. The features of concern that are discussed in Chapter 2 are:

- Skew

- Bimodality

- Outliers

For categorical data, a simple frequency distribution and/or its corresponding bar chart is appropriate. For quantitative data, the single best method of displaying the data is the stemplot, although frequency distributions and their corresponding histogram or polygon are also effective. The stemplot has two major advantages:

- It preserves the raw data whereas the use of class intervals loses all information about differences among scores within each interval.

- It provides both numerical and graphical information in a single figure.

2. Describing data numerically. In describing the results of a scientific study, one of the first tasks is to report numerical descriptions of the data. Such descriptions should include:

- Sample size

- Measure (or measures) of typicality or central tendency

- Measure of dispersion or variability (relevant only for quantitative variables)

For categorical data, the mode is the only appropriate measure of typicality. For quantitative data, the mean is the most commonly used measure of central tendency. If a distribution is symmetrical, then the mean and median will have the same value. However, if the distribution is skewed, this will not be the case and the median may provide a more valid indication of central tendency. The mean is strongly influenced by skew and especially by outliers, whereas the median is a more resistant statistic. It may be appropriate in some cases to report both the mean and the median. In the case of a strongly bimodal distribution, neither the mean nor the median will be appropriate because neither statistic will capture typicality.

The dispersion of a distribution can be described in terms of the location of the first and third quartiles and the distance between them—the interquartile range, *IQR*. These values, along with the median, the minimum, and the maximum score provide a five-point summary of the distribution and can be presented graphically in the form of a box-and-whisker plot (boxplot for short).

The variance and the standard deviation are the measures of dispersion that will be employed for most of the methods of statistical data analysis to be described in later chapters. Unlike a resistant statistic such as the *IQR*, these measures are strongly influenced by outliers.

3. Transforming data (standard scores and linear transformations). It is often desirable to transform data from their initial mean and standard deviation to new values that are more useful or meaningful. There are two basic purposes for such a transformation.

■ Communication. It may be desirable to rescale scores so that they have a mean and standard deviation with values that are easily handled. IQ scores provide the classic example of such scaling. IQ test scores are rescaled to have a mean of 100 and a standard deviation of 15.0. Most other scales such as the SAT are also rescaled to a convenient mean and standard deviation.

■ Comparability. Scores from distributions with different means and standard deviations can be more readily compared if the distributions are rescaled to have the same mean and standard deviation. A common rescaling is to transform raw scores into standard scores (*z*-scores). In such cases the rescaling is to a distribution with a mean of zero and a standard deviation of 1.0.

What You Should Know and Be Able to Do

■ Decide whether a data set has a categorical or quantitative variable and choose an appropriate form for the graphical display of the data. For a categorical variable use a bar chart. For a quantitative variable use a histogram or polygon (either of frequencies or proportions), or a stemplot.

■ Construct the appropriate display. For a bar chart, simply tally the frequencies for each category to determine the height of each bar. For histogram or polygon, first decide on the appropriate number of class intervals (see Section 2.1.2 for rules of thumb). Then tally the frequencies for each interval to determine the height of each bar above the class interval (histogram) or the height of the point above the midpoint of the class interval (polygon).

For the stemplot, first decide on the stem values. These can be determined using the same rules of thumb as for class intervals. Stem values are usually listed from the lowest at the top to the highest at the bottom, but it is not wrong, and in some ways more intuitive, to follow the opposite order. Remember, however, turning the stemplot on its side to gain the appearance of a histogram requires the former ordering of the stem values.

■ Use a histogram, polygon, or stemplot to detect skew, bimodality, or outliers.

■ For a categorical variable, identify the mode.

■ For a quantitative variable:

Calculate the mean, although normally a computer would perform this arithmetic chore.

Locate the median and quartiles. If this is to be done without computer assistance, the stemplot provides the easiest method.

Calculate the interquartile range, *IQR*. You should then know how to use this value to define outliers.

Know the effect of skew and outliers on the relative location of the mean and median.

Calculate the variance and standard deviation, although normally a computer would perform this arithmetic chore.

Form a rough estimate of the mean and standard deviation by inspecting a data set.

Know the effect of outliers on the variance.

Draw a boxplot given the values of Q_1, Md, and Q_3.

State the effect of a linear transformation on the mean and standard deviation.

Rescale scores to

Standard (*z*) scores.

A distribution with any specified mean and standard deviation.

Short-Answer Questions

1. What is the difference between a histogram and a bar chart with respect to

   a. appearance?

   b. type of data to which each is applicable?

   c. ordering of values of the predictor variable on the x-axis?

2. State a rule of thumb for deciding on the number of class intervals for a histogram, frequency polygon, or stemplot. Does the number of observations play any role in this decision?

3. Explain the distinction between the stem and the leaf of a stemplot.

4. Why is it meaningless to describe the frequency distribution (bar chart) of a categorical variable as being skewed?

5. Define deciles and percentiles. Given that "quint" means five, what do you suppose a *quintile* is?

6. Explain why the median is described as a resistant statistic, but the mean is not. Does the same explanation hold for the claim that the *IQR* is a resistant measure of spread whereas the variance is not?

7. Explain the "zero sum principle."

8. Why is the variance also called a mean square? In forming a variance, what values are being averaged?

9. Explain in words how a boxplot is constructed.

10. What are the mean and variance of a distribution of standard scores?

11. Is it possible for a distribution of standard scores be skewed or bimodal?

12. Jane obtains a score of 35 out of 40 in a class quiz. Joan scored 31 on the same test. The instructor transforms the scores so that they have a different mean and standard deviation. Explain why, after this transformation, Jane's new score will remain higher than Joan's new score.

Short Problems

Problem 1 A sample of 20 measurements of height has a mean of 68 inches, a median of 67 inches and a standard deviation of 3.5 inches. The experimenter decides to convert these scores from inches to centimeters. Calculate the mean, median, and standard deviation of these rescaled measures (1 inch = 2.5 cm).

Problem 2 Listed below are the weights (in pounds) of 20 members of a school football squad. If you wanted to impress an opposing team with the high overall weight of your squad, which statistic would you use, the mean, median, or standard deviation?

180	181	179	240	183	181	178	230	182	179
184	221	180	183	180	181	180	232	183	182.

Problem 3 An instructor in a large course gives a long multiple-choice exam that produces a class mean of 122 and a standard deviation of 14.2. The instructor decides to rescale these scores so that they have a mean of 70 and a standard deviation of 10. A student scored 130 on the original test. What is the value of her rescaled score? Give your answer to the nearest whole number.

Problem 4 If the mean of a distribution is greater than the median, what feature of this distribution might an experimenter worry about? If, despite a large sample, very few scores are in the vicinity of the mean or median, what feature of this distribution might an experimenter worry about?

Problem 5 Decide whether each of the following statements is true or false.

a. In a distribution with positive skew, the median could be greater than the third quartile, Q_3.

b. Symmetrical distributions cannot be bimodal.

c. In a symmetrical distribution, the median equals the mean.

d. The interval between Q_1 and the median need not equal the interval between the median and Q_3.

e. If the scores in a distribution are divided by 2, the *IQR* is unchanged because *both* Q_1 and Q_3 are divided by 2.

f. If the scores in a distribution are divided by 2, the mean and the standard deviation are divided by 2.

g. If the scores in a distribution are multiplied by 3, the mean is multiplied by 3 but the median and mode remain unchanged.

h. Outliers at the low end of the distribution greatly reduce the variance, whereas outliers at the high end of the distribution greatly increase the variance.

i. If a distribution is negatively skewed, the median will usually be greater than the mean.

Exercises in the Analysis of Realistic Data

The purpose of the following exercises is to apply the procedures described in Chapter 2. The general strategy will be to display the data in the form of a stemplot, inspect the distribution for any problems such as severe skew or outliers, and then obtain basic descriptive statistics such as the mean and standard deviation. Once these statistics have been obtained, scores can be linearly transformed to standard scores or to a distribution with any specified mean and standard deviation.

Students using computer software should obtain a stemplot and basic statistics using the data sets provided. Part I of the SAS and SPSS instructional modules explains how to import data. Students not using computers can make use of the "results of basic computations" that accompany each problem to avoid most of the heavy arithmetic.

Set 1: Understanding Memory

Exercise 1. Short-term forgetting (Data Set MEM_01.dat)

A first step toward understanding why our memory sometimes fails is to realize that, very often, a great deal of forgetting occurs almost immediately after an event takes place. We have all experienced the embarrassment of being introduced to someone, only to realize a mere 30 seconds later that we have completely forgotten the person's name. Such forgetting is not at all unusual; its major cause is that after hearing the name we are immediately distracted by the demands of conversation. It is rather like looking up a telephone number, but before dialing being interrupted by a roommate with a series of questions. By the time the questions are answered the telephone number is forgotten. The time course of such forgetting is easily demonstrated and was the subject of classic experiments completed in the late 1950s. These experiments serve as a model for the first data set.

In this short-term memory experiment (described briefly in Exercise 5 at the end of Chapter 1) participants were shown three consonants; for example P-S-K. Immediately upon reading the three letters, a three-digit number appears (e.g. 708). Participants were instructed that upon seeing the number, they were to begin immediately counting backwards by threes: "705, 702, . . . " as rapidly as possible until a tone sounded, at which point they were to recall the initial three consonants.

Participants were given a long sequence of these trials. The tone occurred after 3, 6, 9, 12, 15, or 18 seconds of backward counting. These were the six different retention intervals. This experimental procedure is commonly referred to as the Brown-Peterson paradigm and is named after the previously mentioned researchers who first used it to study forgetting in short-term memory. The short-term memory of each of the 35 participants is the proportion of correctly remembered items at each retention interval.

The resulting data are shown in the following table and given in the *ASCII Files*.

Data Set MEM_01.dat

RETENTION INTERVAL

Participant	3 sec	6 sec	9 sec	12 sec	15 sec	18 sec
1	.58	.60	.32	.08	.16	.16
2	.80	.28	.24	.24	.12	.08
3	.90	.68	.34	.36	.16	.20
4	.90	.72	.40	.36	.16	.16
5	.48	.32	.34	.00	.00	.08
6	.84	.28	.24	.20	.16	.04
7	.56	.44	.28	.32	.12	.08
8	.90	.40	.32	.32	.08	.00
9	.90	.52	.36	.20	.12	.08
10	.48	.40	.20	.16	.00	.00
11	.98	.56	.56	.48	.26	.24
12	.84	.68	.44	.16	.16	.08
13	.84	.32	.20	.16	.24	.04
14	.80	.64	.26	.32	.16	.28
15	.72	.28	.32	.36	.00	.00
16	.96	.64	.40	.36	.12	.24
17	.84	.36	.28	.12	.08	.04
18	.64	.16	.44	.20	.16	.00
19	.80	.52	.44	.20	.24	.16
20	.96	.64	.52	.36	.24	.20

Participant	3 sec	6 sec	9 sec	12 sec	15 sec	18 sec
21	.88	.52	.26	.20	.08	.00
22	.84	.32	.40	.24	.04	.04
23	.76	.28	.32	.08	.16	.04
24	.92	.52	.36	.32	.20	.16
25	.76	.32	.36	.20	.00	.00
26	.80	.48	.36	.36	.08	.08
27	.88	.52	.36	.28	.24	.08
28	.96	.68	.48	.24	.26	.12
29	.64	.68	.26	.12	.00	.16
30	.68	.60	.36	.28	.16	.08
31	.88	.44	.32	.24	.12	.16
32	.96	.72	.58	.28	.20	.12
33	.64	.56	.36	.24	.12	.04
34	.68	.40	.28	.16	.04	.00
35	.88	.68	.36	.24	.08	.16
Mean	.80	.49	.35	.24	.13	.10
SD	.14	.16	.09	.10	.08	.08

1. Using the means for each retention interval, plot a graph of the forgetting curve, setting out the values of the retention interval on the x-axis and the mean proportion of items recalled on the y-axis.

2. As practice at constructing stemplots, choose any one of the retention intervals and construct a stemplot for the data for that interval. Check your plot against the one shown below. For any particular plot you may have decided on a different stem width. One convenient feature of a stemplot is that it is easy to convert from one stem width to another by breaking up or combining leaves. If you chose a different stem width you can still use the following stemplots to check your own.

3. Inspect the stemplots. Is there any evidence of outliers, bimodality or skew?

4. The stemplot for the 3-sec retention interval indicates negative skew whereas the 18-sec retention interval indicates positive skew. Can you suggest a reason for this pattern?

3 sec		6 sec		9 sec	
.4	88	.1	6	.2	0044
.5	68	.2	888	.2	666888
.6	44488	.3	22226	.3	2222244
.7	266	.4	000448	.3	66666666

3 sec *(continued)*		6 sec *(continued)*		9 sec *(continued)*	
.8	0000444448888	.5	2222266	.4	000444
.9	0000266668	.6	0044488888	.4	8
		.7	22	.5	2
				.5	68

12 sec		15 sec		18 sec	
.0	0	.0	0000044	.0	0000000444444
.0	88	.0	88888	.0	88888888
.1	22	.1	222222	.1	22
.1	6666	.1	666666666	.1	6666666
.2	00000444444	.2	004444	.2	0044
.2	888	.2	66	.2	8
.3	2222				
.3	666666				
.4					
.4	8				

Why does forgetting occur so rapidly? What can be done to prevent it happening? In the case of trying to remember names, most manuals on memory improvement advise that you try to associate the person's name to something more meaningful. The general principle that closely ties effective remembering to the extraction and elaboration of meaning is the foundation of most mnemonic strategies found in manuals on improving memory. The following exercises illustrate the effectiveness of these strategies.

Exercise 2. Generation effect (Data Set MEM_02.dat)

One way of trying to remember material is to read it over and over. Is there a better way? The following experiment, described in Exercise 1 of Chapter 1, suggests such a way.

Recall from Chapter 1 that a total of 50 participants took part in this experiment, 25 being randomly assigned to each of two conditions. Participants in the "read" condition saw a list of 40 common words such as **small** and **happy** and were instructed to read them out loud in preparation for a subsequent recall test. Participants in the "generate" condition saw a list of words along with the first letter of a second word. They were instructed to generate a word that started with this letter and meant the opposite of that word. For example, if they saw **large s** ___, they responded "small" or if they saw **sad h___** they responded "happy." In this way both groups spoke an identical set of words, the "read" groups by reading them directly, the generate group by generating the target word in response to the cue word.

Following the presentation of the words, all participants were asked to recall as many of the spoken words as they could remember. The number of words recalled by each participant is shown in data set MEM_02.dat.

Data Set MEM_02.dat

Read

10	7	9	12	10	11	12	8	14	8	13	6	8
13	11	9	12	9	7	10	7	7	10	5	10	

Generate:

12	13	16	9	16	18	18	17	17	19	14	10	9
14	14	16	13	16	10	14	15	16	14	19	14	

1. Construct a stemplot for the data in each condition. Note that for these data the stems could be whole numbers rather than intervals. In such cases the leaves are simply frequency counts of each score. Such tallies are commonly denoted by 0.

2. Locate the median, Q_1 and Q_3 in the stemplot. Check your location against the values shown in the following table.

3. Fill in the missing entries in the following table.

	n	Mean	SS	Var.	SD	Q_1	Md	Q_3	IQR
Read	——	9.5	134.2	——	——	7.5	10.0	11.5	——
Generate	——	14.5	202.2	——	——	13.0	14.0	16.5	——

4. Do the data for either condition contain outliers?

5. Draw a boxplot for each condition.

Exercise 3. Memory for chess positions (Data Set MEM_03.dat)

If memory can be improved by making meaningful connections to existing knowledge, then expertise in a given area should enhance memory for information within that area. A classic example of this fact is memory for chess positions. For chess experts, chess positions should be more meaningful than they are for novices.

   This study compared the memory of chess experts for chess positions with the memory of novices for the same positions. Two groups were selected. Group A consisted of 22 chess experts, Group B 22 chess novices. The response measure was the number of board positions correctly recalled. The results were as follows.

Data Set MEM_03.dat

Experts

9	9	12	12	18	7	4	6	12	8	13
11	21	6	12	10	16	14	13	17	15	16

Novices

5	8	13	11	6	11	3	5	7	5	10
10	9	9	6	13	9	4	4	7	7	5

1. Construct a stemplot for the data in each condition. Note that for these data (as in Exercise 2) the stems can be whole numbers rather than intervals.

2. Fill in the missing entries in the following table.

	n	Mean	SS	Var.	SD
Experts	22	11.86	388.6	——	——
Novices	22	7.59	179.3	——	——

3. Do the data for either condition appear to have outliers?

4. Draw a boxplot for each condition.

5. Is the predictor variable in this experiment natural or manipulated? What implications does your answer have for the interpretation of the results from this experiment?

Exercise 4. Memory for random chess positions (Data Set MEM_04.dat)

A possible interpretation of the chess experiment in Investigation 3, is that chess experts have naturally better memories to start with, and that rather than being the *result* of their chess experience, their superior memories is a major cause of their becoming expert.

To explore this possibility, a second Investigation is conducted, again comparing the memory of expert and novice chess players. There were 28 participants in each group. In the original investigation, the board positions to be remembered were meaningful game positions. In this second investigation the chess pieces were arranged in random positions on the board. As in the first investigation, the response measure was the number of board positions correctly recalled. The results are given in data set MEM_04.dat.

Data Set MEM_04.dat

Expert

9	4	7	9	8	11	10	11	10	9	9	7	9	9
8	8	6	5	9	8	10	5	6	3	4	5	9	9

Novice

3	8	6	8	11	3	8	5	8	8	10	10	5	7
8	7	2	8	6	6	7	3	11	6	4	9	8	9

1. Construct a stemplot for the data in each condition.

2. Fill in the missing entries in the following table.

	n	Mean	SS	Var.	SD	Q_1	Md	Q_3	IQR
Experts	28	7.75	——	4.81	——	6.00	8.50	9.00	——
Novices	28	6.93	——	5.92	——	5.25	7.50	8.00	——

3. Do the data for either condition appear to contain outliers?

4. Draw a boxplot for each condition.

Set 2: Child Behavior

Exercise 5. Familiarity and compliance (Data Set CHILD_1.dat)

Are children more likely to comply promptly to a request made by a parent or to one made by a stranger? In this investigation, seven-year-old children are taken to a laboratory playroom that has intentionally been left messy. After entering the room the child is asked to clean it up. In one condition the child's parent makes the request. In a second condition the request is made by a stranger (the investigator).

A total of 50 seven-year-old children participated in the experiment. They were assigned randomly to the conditions, 25 in each. The response measure was the time taken (latency in seconds) for the child to begin the task.

Data Set CHILD_1.dat

Parent

24 25 34 13 36 42 40 39 38 43 27 10 12
28 30 35 24 34 15 29 31 36 30 46 28

Stranger

20 11 18 27 20 35 33 15 28 16 22 5 17
22 33 18 28 19 12 22 13 14 22 4 23

1. The investigator in this study would like to decide whether there is any difference between the two conditions. In order to proceed it is important to establish that the data contain no abnormal features. What steps should the investigator take?

2. Perform whatever steps are necessary to establish that the data contain no abnormalities.

3. The mean for one condition is 19.9 and for the other it is 30.0. By inspecting the data, judge which mean belongs to which condition.

4. The sum of squares for the parent-request group was 2297.0 and for the stranger-request group is was 1534.6. Calculate the standard deviation for each condition.

Exercise 6. Perspective taking in parents (Data Set CHILD_2.dat)

Parents are frequently faced with the need to decide whether or not to punish their child for misbehavior. What factors influence this decision? An obvious factor is the nature of the misbehavior. Another possible influence is the extent to which parents view the misbehavior from their own perspective or from that of the child. This study investigated this possibility using three conditions. A total of 45 mothers participated. They were assigned randomly to the three conditions, 15 in each. All mothers read a story about a child's misbehavior.

The mothers in Condition 1 were instructed to "think about the misbehavior from the child's point of view. What was the child thinking and feeling at the time the behavior occurred?"

The mothers in Condition 2 were instructed to "think about the misbehavior from the your own point of view. What would you be thinking and feeling at the time the behavior occurred?"

The mothers in Condition 3 formed a control condition and were given no special instructions.

After reading the story all parents used a seven-point scale to rate the likelihood that they would punish the child for the misdeed.

Data Set CHILD_2.dat

Condition 1 "Think about the misbehavior from the child's point of view."

2	1	2	3	2	5	4	1	3	5	2	1	2	2	4

Condition 2 "Think about the misbehavior from your point of view."

4	5	4	3	4	3	3	4	2	6	3	3	4	6	5

Condition 3 Control condition

6	3	5	5	6	4	2	3	4	4	5	3	4	2	4

1. Calculate the mean for each condition.

2. The sum of squares for the three conditions are 25.6, 18.9, and 22.0. Calculate the standard deviation for each condition.

Exercise 7. Authoritarianism (Data Set CHILD_3.dat)

In order to investigate the relationship between the authoritarianism of parents and various characteristics of their children, a researcher constructs a scale to measure authoritarianism, consisting of true/false questions. When tested on a random sample of 150 mothers, the scale produces scores ranging from 15 (low authoritarianism) to 75 (high authoritarianism) with a mean of 45.7 and a standard deviation of 9.1.

The researcher decides that, for ease of reporting, the scale should be transformed to have a mean of 50 and a standard deviation of 10. Describe how the researcher would accomplish this aim. If a parent scored 22 on the original scale, what would be the transformed score?

Having constructed this scale, the first question asked is whether the academic achievement of the child is related to the degree of authoritarianism of the mother. To address this question, a sample of 12-year-olds ($n = 65$) is chosen and their academic performance is measured in terms of their average course grade, expressed as a percentage. An authoritarianism score is obtained for each child's mother.

Data Set CHILD_3.dat

Auth.	Grade	Auth.	Grade	Auth.	Grade
50	74	52	69	37	76
41	75	34	80	51	82
48	61	53	71	56	67
57	62	44	64	50	64
50	76	45	76	66	68
68	58	54	67	48	69
63	64	33	75	50	70
45	67	56	63	39	72
58	63	62	64	53	69
46	78	60	63	54	74

Auth.	Grade	Auth.	Grade	Auth.	Grade
52	55	59	61	47	55
29	67	58	64	42	86
47	76	63	56	48	70
52	68	47	61	59	54
63	63	30	80	56	61
48	75	32	61	53	75
58	67	48	71	27	70
49	56	38	62	26	67
42	81	55	70	58	59
52	61	44	74	40	69
43	63	54	73	49	78
44	67	35	77		

1. Examine the stemplot for each measure. Locate the median, Q_1, and Q_3 for each distribution and check your answers against the values given below.

Authoritarianism

2	6
2	79
3	0234
3	5789
4	01223444
4	5567778888899
5	000012222333444
5	56667888899
6	02333
6	68

School grades

5	4
5	556689
6	111111223333334444
6	7777777889999
7	00001123444
7	55556666788
8	0012
8	6

2. Fill in the missing entries in the following table.

	n	Mean	SD	Q_1	Md	Q_3	IQR
Mother's authoritarianism	65	48.8	9.67	43.5	50.0	56.0	——
Child's school grade	65	68.2	7.37	63.0	68.0	74.5	——

3. Does the distribution for either condition contain outliers?

4. Draw a boxplot for each condition.

Exercise 8. Household chores and children's concern for others (Data Set CHILD_4.dat)

Many parents would like to encourage their children to develop a spontaneous concern for the well-being of others. How might such a concern be fostered? One possibility is that concern for others is related to the performance of household chores. The following study was conducted to explore this relationship.

A total of 60 children aged 12 to 14 years participated in the study. Estimates were obtained of the amount of time the children spent each week on various forms of household chores. One of these forms was "family routine" chores such as setting the table, or taking out garbage. These chores are performed routinely in the sense that the child does them without being explicitly requested each time.

Each child was also observed for a total of 30 hours over a three-month period and any act of spontaneous concern for others was recorded. The number of such acts formed the child's "concern" score. The "family routine chores" and the "concern-for-others" score for each child are found in data set CHILD_4.dat, which records the time in minutes spent performing family routine chores (Cho) and the number of spontaneous expressions of concern for others (Con) from a sample of 60 children.

Data Set CHILD_4.dat

Cho	Con	Cho	Con	Cho	Con
24	21	10	13	29	25
15	19	18	7	14	25
22	24	26	20	28	26
31	29	8	5	9	7
24	18	27	14	23	15
42	26	18	20	25	22
37	27	12	29	30	25
19	33	28	9	24	20
32	14	7	15	40	19
20	21	30	13	22	18
26	20	36	27	24	18
5	1	34	15	23	16
21	34	33	16	27	15
26	14	32	30	28	30
37	32	37	27	21	21
22	4	21	12	16	9
32	19	4	13	22	21
23	30	6	8	33	37
16	19	22	20	30	23
26	25	24	20	27	19

1. Check the following stemplots for any evidence of abnormalities in the data.

2. Use the stemplots to locate the median, Q_1, and Q_2 for each distribution check your result with the value in the following table.

Chores

0	4
0	56789
1	024
1	566889
2	01112222233344444
2	566667778889
3	0001222334
3	6777
4	02

Concern

0	14
0	577899
1	2333444
1	55556688899999
2	0000001111234
2	55556677799
3	000234
3	7

3. Fill in the missing entries in the following table.

	n	Mean	SS	SD	Q_1	Md	Q_3	IQR
Family routine chores	60	23.8	4623.6	——	19.5	24.0	30.0	——
Concern for others	60	19.6	3634.8	——	14.5	20.0	25.0	——

4. Someone inspecting the data claims that child 4 (Cho = 31 and Con = 29) in the data set is an example that supports the claim that children who perform a lot of family routine chores show a great deal of concern for others. How might this claim for child 4 be supported quantitatively? Why does this example not provide convincing evidence for the general conclusion that children who perform a lot of family routine chores show a great deal of concern for others?

Set 3: Test Construction and Evaluation

Exercise 9. Developing an aptitude test (Data Set TEST_1.dat)

Imagine you are the Director of Psychological Services for a large school board. The board decides it would like to use its own scholastic aptitude test so that they will have a test more appropriate to local circumstances. You are charged with the responsibility of constructing and evaluating the test.

After preliminary testing you settle on a test with 160 items. Each item is scored 0 (incorrect) or 1 (correct). Total test scores can therefore range from 0 to 160. In order to conduct a final evaluation, the test is administered to a sample of 200 students. The stemplot for the resulting distribution of scores is shown below.

2	289
3	02335799
4	12233578
5	0011444555668899
6	000001111333445577899999
7	11122222333334667788888
8	001112224444556666777779
9	000000113333355777888999999
10	00000022222333344666788
11	0000000236666779999
12	00035599
13	0222333344
14	0129
15	125

1. Locate the median, Q_1 and Q_3 in the stemplot. Check your location against the values shown in the following table.

2. Fill in the missing entries in the following table.

n	Mean	SS	Var.	SD	Q_1	Md	Q_3	IQR
200	86.36	160,862	——	——	65	86	104	——

3. Do the data contain outliers?

4. Draw a boxplot for each condition.

5. You decide to rescale the scores on your aptitude test to a more convenient form—a scale with a mean of 50.0 and a standard deviation of 10.0. To facilitate the recording of the rescaled score you develop a table that gives the rescaled score corresponding to every original score. What would the tabled entries be for original scores of 70, 80, and 90?

Exercise 10. Extroversion scale (Data Set TEST_2.dat)

A personality researcher decides to construct a scale to measure extroversion. The scale consists of 25 items to which participants respond by indicating on a five-point scale (1–5) the extent to which they agree with the statement. The score on the scale is the sum of the ranks across the 25 items. Thus a participant's score on the scale can range from 25 (extreme introversion) to 125 (extreme extroversion).

As part of the valuation of the scale, the researcher administers the scale to 120 volunteers.

93	34	56	61	54	57	69	55	87	51	57	49
68	75	84	73	58	58	60	28	73	64	56	60
30	64	43	62	37	47	83	47	43	62	55	66
77	43	65	54	62	65	71	73	54	60	65	94
40	66	58	36	94	44	75	26	49	82	63	69
51	37	57	41	44	63	83	32	62	39	65	40
44	75	64	45	61	49	62	60	67	76	69	44

```
50  63  66  60  45  58  54  52  45  79  65  43
65  89  63  55  43  87  56  65  56  71  72  67
54  81  84  53  55  62  37  52  86  50  46  53
```

1. Construct a stemplot for these data and check that the distribution is approximately symmetrical and unimodal.

2. Locate the median, Q_1 and Q_3 in the stemplot.

3. Construct a boxplot for the data and calculate the *IQR* of the distribution.

3 MODELING DATA AND THE ESTIMATION OF PARAMETERS

Overview

Chapter 3 develops in greater detail the basic concepts, introduced in Chapter 1, of stating a model, fitting it to data, and obtaining a measure of the goodness of the fit.

1. A model for replicability. The simplest of all models is one that expresses the principle of strict replicability. This principle states that for a fixed condition all observations should be the same. A model that captures this principle is therefore $\hat{Y} = \mu$, where $\hat{Y}$ is the predicted observation and μ is a constant. In the model for replicability, the constant μ is called the parameter of the model. Parameters are the constants that enter into the models prediction rule. All models have one or more parameters.

With real data, models rarely make perfect predictions. In the case of models for replicability, such imperfect predictions are reflected in any variability among observations. The difference between the observation (Y) and the model's prediction ($\hat{Y}$) is called a residual, e. This relationship between the model's prediction and data can be summarized in the following statement.

$$\text{observation (Y)} = \text{model prediction } (\hat{Y}) + \text{residual } (e)$$

or

$$Y = \hat{Y} + e = \mu + e$$

Because the parameters of a model are unknown, their values must be estimated from the data. These point estimates can then be used to fit the model to the data. The prediction rule generates fitted values so we can describe the data as

$$\text{observation} = \text{model fit} + \text{residual}$$

A goal in fitting the model is to obtain estimates of its parameters that make the residuals as small as possible. The sum of squares of the residuals (SS_e) provides a measure of the overall size of the residuals and thus is an index of goodness of fit. The larger the sum of squares, the poorer the fit. A desirable point estimate of a parameter is therefore one that yields the smallest possible sum of squares. Such an estimate is said to satisfy the criterion of least squares. In the model for replicability, the least squares estimate of the parameter, μ, is the sample mean, $\bar{Y}$. In this case, the fitted model is therefore

$$Y = \bar{Y} + e.$$

If data are thought of as a sample from a larger population of potential observations, then the parameters of the model can be thought of as the population counterpart of the sample statistic. Thus the parameter μ is interpreted as the population mean—the population counterpart of the sample mean, $\overline{Y}$. According to this view a model is a statement about what is true in the population.

Once the model has been fitted, the sum of squared residuals (SS_e) can be calculated as a measure of the goodness of fit of the model. Dividing SS_e by its degrees of freedom (n - 1) gives the mean square of the residuals, MS_e.

2. Larger models. Models for studies with a categorical predictor variable that has two or more values (conditions) are simple extensions of the model for replicability of a single condition. A full model for such experiments has one parameter for each condition. Each parameter corresponds to the population mean for that condition. Under the full model, the sum of squares of the residuals is SS_e, obtained by adding up the sum of squared residuals for each condition. The corresponding mean square (MS_e) is obtained by dividing this sum of squares by degrees of freedom. The calculation of a single value of MS_e to represent two or more conditions is based on the assumption that the population variances are the same for both conditions: the assumption of homogeneity of variance.

The null model is a simpler model based on the null hypothesis that the parameters (the population means for each of the conditions) are all equal. Under the null model, the sum of squares of the residuals is simply the sum of squares for all the scores and is denoted by SS_{total}.

The difference between SS_{total} and SS_e quantifies the variability accounted for by the full model. This difference is denoted SS_{model}. The proportion of variance accounted for by the full model can be quantified by expressing SS_{model} as a proportion of SS_{total}. This proportion is denoted by R^2.

$$R^2 = \frac{SS_{model}}{SS_{total}}$$

3. Effect size. A commonly used measure of effect size is Cohen's **d**. Its estimate is

$$\hat{d} = \frac{\overline{Y}_1 - \overline{Y}_2}{\sqrt{MS_e}}$$

This measure is especially useful in comparing effect sizes across different experiments.

4. Models for quantitative predictor variables. Models for categorical predictor variables differ from those for quantitative predictor variables. Models for categorical predictor variable predict values of the response variable only for values of the predictor variable included in the experiment. Models for quantitative predictor variables, on the other hand, consist of a rule that predicts values of the response variable for any value of the predictor variable within a specified range, whether or not that value was a condition in the experiment.

What You Should Know and Be Able to Do

■ Explain why the model $Y = \mu + e$ is a model of the principle of strict replicability, including a description of each component (Y, μ, and e) of the model.

■ Explain what is meant by the statement "μ is a parameter of the model."

■ Explain the meaning of the expression "the sample mean is the least squares estimate of the parameter μ" and why μ can be interpreted as the population mean.

■ Know the formula for SS_e. Reconstruct this formula, not through rote memorization, but through your understanding of the meaning of SS_e.

■ Explain how the size of the residuals (measured by SS_e or MS_e) is a measure of how well the model fits the data.

■ Understand how the model of strict replicability for a single condition can be extended to cover two or more conditions and how to fit such a model to the data.

■ Given the sum of squares for each condition obtain a single estimate of the mean square of the residuals.

■ Explain the distinction between a full and a null model.

■ Explain why the sum of squares of the residuals for the null model is also called SS_{total}. You should be able to reconstruct the formula for SS_{total}, not through rote memorization, but through your understanding of the meaning of SS_{total}.

■ Explain the meaning of SS_{model}.

■ Define and explain the meaning of R^2.

■ Define and explain the meaning of the effect size, **d**.

■ Describe the difference between models for quantitative and categorical predictor variables.

Short-Answer Questions

1. Explain the difference in meaning between the symbols $\hat{Y}$ and Y.

2. State and explain the four steps in fitting a model to data.

3. In the model $Y = \mu + e$, what component can be thought of as analogous to a signal and what component can be thought of as analogous to noise?

4. Why is it usually necessary to estimate the parameters of a model?

5. Why can we be absolutely certain that no other estimate of μ will result in a smaller sum of squares than that obtained using $\overline{Y}$?

6. Why must the value of SS_e always be positive?

7. Why must the value of R^2 always be positive?

8. Explain what is meant by the assumption of homogeneity of variance. Give one reason why it is important.

9. What is the relationship between SS_e, SS_{total}, SS_{model}?

10. In what sense is the rationale for calculating the effect size, d, similar to that for calculating standard (z) scores?

Short Problems

Problem 1 Five observations from a single condition are as follows:

4   6   5   8   2

Following the four steps described in Section 3.1.2, fit the model $Y_1 = \mu_1 + e$ to these data, including the calculation of SS_e and MS_e. The subscript has been added to μ in anticipation of Problems 2 and 3 that follow.

Problem 2 Five observations from a different condition from that of Problem 1 (but part of the same experiment) are as follows:

1   4   6   1   3

Following the four steps described in Section 3.1.2, fit the model $Y_2 = \mu_2 + e$ to these data, including the calculation of SS_e for these five observations.

Problem 3 Now calculate the value of SS_e and MS_e for the full model for the experiment as a whole. On what assumption is the calculation of this overall value of MS_e based?

Problem 4 Treating the two sets of data in Problems 1 and 2 as observations from two conditions of a single experiment, follow the four steps described in Section 3.1.2 and fit the null model ($Y = m + e$) to the data, including the calculation of SS_{total}.

Problem 5 Calculate R^2 and an estimate of **d** for these data.

Problem 6 Suppose the data in Problem 1 are replaced by the following observations:

6   8   7   10   4

Notice that these new observations were obtained by simply adding 2 to each of the original observations. Why does this leave the value of SS_e unchanged? Will the values of SS_{total}, d, and R^2 be unchanged? Check your answer by recalculating the values of each of these statistics.

Exercises in the Analysis of Realistic Data

Set 1: Understanding Memory

Exercise 1. Generation effect (Data Set MEM_02.dat)

A total of 50 participants took part in this experiment, 25 being randomly assigned to each of two conditions: The "read" condition and the "generate" condition. Following the presentation of the words, all participants were asked to recall as many of the spoken words as they could remember. The number of words recalled by each participant were shown in data set MEM_02.dat (Chapter 2, Exercise 2).

The full model for this experiment is:

Read Condition: $Y_1 = \mu_1 + e$

$$\text{Generate Condition:} \quad Y_2 = \mu_2 + e$$

1. Using the results given in Exercise 2 of Chapter 2, write down the estimates of μ_1 and μ_2.

2. The sums of squares were $SS_1 = 134.24$ and $SS_2 = 202.24$. Use these values to calculate the combined sum of squares (SS_e) and then the overall mean square (MS_e) of these residuals across the two conditions.

3. Calculate an estimate of Cohen's **d** statistic as a measure of effect size in this experiment.

4. Write a model for this experiment under the null hypothesis that $\mu_1 = \mu_2 = \mu$. The sum of squares of the residuals under this simpler null hypothesis model is $SS_{total} = 648.98$. Calculate the value of SS_{model} and R^2 for this experiment.

Exercise 2. Memory for chess positions (Data Set MEM_03.dat)

This study compared the memory of chess experts for chess positions with the memory of novices for the same positions. Two groups were selected. Group A consisted of 22 chess experts, Group B of 22 chess novices. The response measure was the number of board positions correctly recalled.

1. Write out the full model and the null model for this experiment and, using the results given in Exercise 3 of Chapter 2, write down the estimates of the various parameters. Using results from Exercise 3 of Chapter 2, calculate the combined sum of squares (SS_e) and then the overall mean square (MS_e) of these residuals across the two conditions.

2. Given the result that $SS_{total} = 768.7$, calculate the value of SS_{model} and R^2 for this experiment.

Set 2: Child Behavior

Exercise 3. Familiarity and compliance (Data Set CHILD_1.dat)

In this investigation, seven-year-old children are taken to a laboratory playroom that has intentionally been left messy. After entering the room the child is asked to clean it up. In one condition the child's parent makes the request. In a second condition the request is made by a stranger (the investigator). A total of 50 seven-year-old children participated in the experiment. They were assigned randomly to the conditions, 25 in each. The response measure was the time taken (latency in seconds) for the child to begin the task.

Using the results given in Exercise 5 of Chapter 2, calculate an estimate of Cohen's d statistic as a measure of effect size in this experiment. Given that $SS_{total} = 5101.68$, calculate the value of R^2.

Exercise 4. Perspective taking in parents (Data Set CHILD_2.dat)

What factors influence a parent's decision whether or not to punish a child for misbehavior? This study used three conditions. A total of 45 mothers participated. They were assigned randomly to the three conditions, 15 in each. All mothers read a story about a child's misbehavior.

The mothers in Condition 1 were instructed to "think about the misbehavior from the child's point of view. What was the child thinking and feeling at the time when the behavior occurred?"

The mothers in Condition 2 were instructed to "think about the misbehavior from your point of view. What would you be thinking and feeling at the time when the behavior occurred?"

The mothers in Condition 3 formed a control condition and were given no special instructions.

After reading the story all parents used a seven-point scale to rate the likelihood that they would punish the child for the misdeed.

1. Write out the full model and the null model for this experiment.

2. Using the results given in Exercise 6 of Chapter 2, write down the estimates of the parameters for both full and null models.

3. Using the results given in Exercise 6 of Chapter 2, calculate the combined sum of squares (SS_e) and then the overall mean square (MS_e) of the residuals across the three conditions.

4. The sum of squares of the residuals under this simpler null hypothesis model is $SS_{total} = 85.24$. Calculate the value of SS_{model} and R^2 for this experiment.

4 PROBABILITY DISTRIBUTIONS

Overview

We have seen that the overall goal of data analysis is to find the simplest (the most parsimonious) model that will account for the data. Chapter 2 set out the basic descriptive statistics that we will need for this task, along with methods of checking data for abnormalities. Chapter 3 provided a description of methods for fitting a model to data and measuring goodness of fit as the sum of squared residuals. The method gave a best fitting model in the sense that the model used estimates of the parameters that produced the minimum sum of squared residuals.

Although this least squares fit is an important achievement, it is purely descriptive and leaves a number of questions unanswered:

How close is $\overline{Y}$ to μ? What can we say about the value of $\overline{Y} - \mu$? All we know at this stage is that $\overline{Y}$ is the best *point* estimate of μ, but an investigator will want to know something about the accuracy of this estimate. A least squares estimate may simply be the best of a bad bunch.

We know that R^2 is a descriptive index of how much better the full model accounts for the data compared to the null model. But does the full model provide a sufficiently better fit to the data to warrant discarding the simpler null model in favor of the full model?

1. Probability and the law of large numbers. The first step in addressing these questions is to understand the apparently chaotic behavior of residuals. Chapter 4 showed that despite the chaotic behavior of single elements (balls in the quincunx provided a mechanical model), the cumulative behavior of many elements is orderly and becomes increasingly stable as the number of elements increases. One aspect of this increasing orderliness is the stability of the sample proportion as described by the law of large numbers. This law states that as the number of observations (the sample size) increases, the average difference between a sample proportion and the theoretical probability (or *population* proportion) becomes smaller.

Another way of expressing the law of large numbers is to say that as the number of observations increases, it becomes increasingly improbable that the difference between the sample proportion and the population proportion (or probability) will exceed any specified value, no matter how small that value is. It is this increasing improbability of large residuals with increasing sample size that justifies the claim that the larger the sample, the more *stable* the estimate. As the sample size becomes infinitely large, the sample proportion becomes the probability. The law of large numbers thus provides us with an interpretation of probability as long-run relative frequency.

Equipped with this interpretation of probability, several terms, some previously used informally, can be given a more formal definition.

■ A *random selection* is a selection for which each potential element has an equal probability of being selected.

■ A *random sample* of a given size is a sample chosen such that each potential sample of that size has an equal probability of being selected.

■ Two events (A and B) are *independent* if the probability that A occurs is unaffected by whether or not B occurs.

2. Probability distributions. A probability distribution is a distribution that specifies the probability of different values of a variable. A variable with an associated probability distribution is called a random variable. Random variables may be categorical or quantitative and, if quantitative, they may be either discrete or continuous.

The *binomial distribution* provides a commonly used example of a discrete probability distribution. This distribution specifies a probability for every possible outcome of a process that consists of n trials on each of which one of two possible outcomes occurs with a constant probability, p. A prototypical example is tossing a coin n times and counting the number of heads. The binomial distribution specifies a probability for every possible number of heads. Probabilities for single outcomes or a range of outcomes for the binomial distribution can be obtained from tables once n and p are known, and the relevant outcomes (values of r) have been specified. The number of girl versus boy children in a family of a given size is one of many everyday situations that are equivalent to this prototypical coin-tossing example.

3. Normal distribution. For the probability distribution of a continuous variable, a probability always refers to an interval of values, not to a point value. The height of the probability curve at a given point is the concentration of probability at that point and is referred to as probability density. As n increases to infinity, the binomial distribution becomes a continuous variable with a symmetrical bell-shaped distribution. Known originally as the law of errors, its more familiar name is the normal distribution. The central limit theorem gives the important result that if the values of a variable are the net result of the addition of many independent contributions, then that variable will have an approximately normal distribution. The approximation improves as the number of contributing components increases.

The symmetry of the normal distribution implies that the mean, median and mode have the same value. The probability of a range of values is the area of the curve covered by that range. Thus $P(a < Y < b)$ is the area under the curve between a and b. It is important to realize that the normal distribution is actually a family of distributions. A normal distribution is defined uniquely by specifying the values of two parameters: the mean, μ, and the standard deviation, σ.

Tables of the normal distribution give areas (probabilities) for a distribution with a mean of 0 and a standard deviation of 1.0 (z-scores). Probabilities for distributions not expressed as z-scores can be obtained by converting the original scores to z-scores. It is often necessary to obtain scores that mark off designated probabilities in a normal distribution. These scores can be obtained by reversing the order of the steps used to calculate probabilities for designated scores.

4. Normally distributed residuals. Most of the methods of data analysis to be described assume that residuals are normally and independently distributed. If we are to use the normal distribution to make probability statements, we will need to know the mean and standard deviation of the distribution. The mean of the residuals will be assumed to be zero. This assumption is equivalent to the assumption that the residuals are unbiased. The standard deviation will usually be unknown so that its value must estimated from the data.

What You Should Know and Be Able to Do

■ State and explain the law of large numbers.

■ Explain the relationship between the law of large numbers and the definition of probability as long run relative frequency.

■ Define and explain the meaning of the following terms: random selection, random sample, probability distribution, random variable, independence.

■ Distinguish between categorical, discrete quantitative, and continuous random variables.

■ Given the values of p and n, use tables of the binomial distribution to establish the probability of a given value (or range of values) of a binomially distributed random variable.

■ Explain why continuous random variables need to introduce the concept of probability density.

■ State the central limit theorem and use it to explain why the distributions of so many natural attributes (such as height and weight) appear to be normally distributed.

■ Given the values of μ and σ, use tables of the normal distribution to establish the probability of a given range of values of a normally distributed random variable.

■ Given the values of μ and σ, use tables of the normal distribution to establish the value (or values) of a normally distributed random variable that marks off a designated probability.

Short-Answer Questions

1. A very large organization has a membership made up of 55% men and 45% women. As part of a survey, a random sample of the membership is to be selected. One plan is to select a sample of size 20; another plan is for a sample of 100. Explain how the law of large numbers predicts that the smaller sample is more likely than the larger sample to have a majority of women.

2. Explain what is meant by the expression "select a random sample" as used in the previous question.

3. Imagine a quincunx with nine rows of pins that has been designed such that when a ball hits a pin it has a $\frac{2}{3}$ chance of falling to the left and a $\frac{1}{3}$ chance of falling to the

right. The collecting bins at the bottom are labeled 0 through 9 from left to right. Describe the probability distribution that specifies the long run proportion of balls in each of these collecting bins.

4. The engineer of the quincunx in the previous example claims that whether the ball deflects left or right in one row is independent of the direction of deflection in any of the preceding rows. Explain what this claim means. Extend your explanation to cover the following claim. "When a basketball player shoots, the success of the shot is independent of the success or failure of the previous shot."

5. An experimenter measures simple reaction time as the time it takes to press a key following the onset of a light. Explain what it means to consider this measurement as a random variable. Could IQ be considered a random variable? Explain your answer. If considered as random variables, would reaction time and IQ be treated as continuous or as discrete random variables?

6. Using as examples the variables *country of birth* and *body weight*, explain the distinction between probability and probability density.

7. In 1846 the Belgian astronomer, Adolphe Quetelet reported measurements of the chest circumference of 5,738 Scottish soldiers, claiming that these measurements were normally distributed. Explain how the central limit theorem makes such a claim plausible.

8. Explain why tables of the normal distribution need to report values for only positive values of z.

Short Problems

Problem 1 The quincunx in Question 3 above had nine rows of pins and had been designed such that when a ball hit a pin it had a $\frac{2}{3}$ chance of falling to the left and a $\frac{1}{3}$ chance of falling to the right. The collecting bins at the bottom were labeled 0 through 9 from left to right. Using the tables of the binomial distribution, write down the probabilities associated with each of the 10 collecting bins. Suppose a ball lands in the bin labeled 4. What is the probability that the next ball will also land in that bin?

Problem 2 In a four-alternative multiple-choice exam, the probability of answering a question correctly by chance is .25. If there are 8 questions, what is the probability that a student would pass the exam (answer 4 or more questions correctly) purely by chance?

Problem 3 Scores on an aptitude test are normally distributed with a mean of $\mu = 60$ and a standard deviation of $\sigma = 10.0$. A Board of Education decides to identify the top 5% and the bottom 10% of students on this test for special attention. What are the two critical aptitude test scores that mark off these percentages?

Problem 4 IQ scores have a normal distribution with a mean of 100.0 and a standard deviation of 15.0. What are the two IQ score values (whole numbers) that mark off the middle 80% of IQs in this distribution, leaving 10% in each tail?

Problem 5 In order to test discrimination learning, a simple experiment is planned consisting of 8 trials. On each trial the subject must discriminate between two alternatives, A and B. Suppose that the subject cannot discriminate at all, and therefore just guesses on each trial. Granted this assumption, using tables of the appropriate probability distribution, calculate the exact probability of the subject

a. being correct on fewer than 4 trials

b. being correct on more than 5 trials.

Note: Guessing means choosing between A and B on each trial with a 0.5 probability of being correct.

Problem 6 IQ scores are normally distributed with a mean of 100.0 and a standard deviation of 15.0.

a. What are the first and third quartiles of the IQ distribution?

b. What IQ score marks off the top 10% of IQs?

c. What is the probability that a randomly selected IQ will be greater than 80.0?

d. What is the probability that a randomly selected IQ will be in the range $(80.0 \leq IQ \leq 100.0)$?

e. "Extreme" IQ scores are defined as those greater than 125 or less than 75. What percentage of the population have extreme IQ scores, so defined?

Problem 7 What are the z-scores that mark off the middle 90% of a normal distribution, leaving 5% in each tail? If the scores from a test have a normal distribution with a mean of $\mu = 50$ and a standard deviation of $\sigma = 13$, what are the two test score values that mark off the middle 90% of test scores in this distribution, leaving 5% in each tail?

Problem 8 Test scores are normally distributed with a mean of 70 and a standard deviation of 14. What are the two test score values (to the nearest whole number) that mark off the first and third quartiles of this distribution?

Exercises in the Analysis of Realistic Data

Test Construction and Evaluation

Exercise 1. Developing an aptitude test (Data Set TEST_1.dat)

In Exercise 9 of Chapter 2 you were asked to imagine yourself as the Director of Psychological Services for a large school board charged with the responsibility of constructing and evaluating a scholastic aptitude test.

Having developed the test, one of its uses in future years will be to assist in the placement of students in classes. For this purpose the school board asks you to report each student's performance as a letter rating A through E. They would like the letter A to be assigned to students scoring in the top 20%, the letter B to students scoring in the next 20%, and so on, ending with the letter E referring to students whose performance is in the bottom 20%. As a psychologist you recognize that what the board is requesting is the quintile rank for each student. Quintiles divide a distribution into five equal areas just as quartiles divide it into four equal areas. Four quintiles are needed to divide a distribution into five equal areas. Thus a student with a score greater than the fourth quintile would be scored A, a student with a score between the third and fourth quintile would be scored B, and so on.

You are willing to assume that scores on your test are normally distributed and so you decide to base your quintiles on the normal distribution. Sketch a normal distribu-

4 PROBABILITY DISTRIBUTIONS ■ 37

tion and mark off the rough location of the quintiles and indicate the areas corresponding to each letter designation, A through E. Then, using the table of the normal distribution, find the four z-scores corresponding to the four quintiles.

In Exercise 9 of Chapter 2 the mean and standard deviation of the sample ($n = 200$) was found to be 86.36 and 28.43 respectively. Using these values as estimates of μ and σ, estimate the test scores corresponding to the four quintiles.

Convert the following test scores to their corresponding letter designations: (a) 60, (b) 85, (c) 100, (d) 120.

Exercise 2. Extroversion scale (Data Set TEST_2.dat)

The researcher would like to classify people into three groups: Extrovert, Introvert, and neither (intermediate between the two). To achieve this purpose the researcher decides to use the extroversion scale described in Exercise 10 of Chapter 2. An extrovert is defined as someone who scores in the top one-third of the population on this scale. An introvert is defined as someone who scores in the bottom one-third of the population on this scale.

The researcher is willing to assume that scores on the extroversion scale are normally distributed and so decides to base the classification boundaries on the normal distribution. Sketch a normal distribution and mark off the rough location of these two boundaries and indicate the areas corresponding to Introvert, Intermediate, and Extrovert. Then, using the table of the normal distribution, find the two z-scores corresponding to the two boundaries.

The mean and standard deviation of the sample ($n = 120$) was found to be 58.97 and 14.22 respectively. Using these values as estimates of μ and σ, estimate the test scores (to the nearest whole number) corresponding to the two boundaries. Classify the participants who obtained the following test scores into their appropriate categories: (a) 50; (b) 60; (c) 70.

5 SAMPLING DISTRIBUTIONS AND INTERVAL ESTIMATION

Overview

1. Sample mean as a random variable. The fundamental concept that underlies everything in Chapter 5 is the idea that the observed sample mean can be considered the value of a variable. This conception requires an act of imagination because in practice only one value of $\bar{Y}$ will be obtained. The first step in understanding how $\bar{Y}$ can be considered a variable is to realize that the sample on which a single value of $\bar{Y}$ was obtained is just one sample from among the infinitely many samples that might have been drawn. What you must imagine is the population of infinitely many $\bar{Y}$ values consisting of means of all possible samples of a given size. Thus the one value of $\bar{Y}$ actually obtained can be thought of as a single observation from a population of $\bar{Y}$ values.

The next step in this development is to treat the obtained sample mean as one value of a *random variable*. Recall that a random variable is simply a variable on which a probability distribution has been defined. The probability distribution that is defined on the variable $\bar{Y}$ is termed the sampling distribution of the mean. The general term sampling distribution is used to describe a probability distribution whenever the random variable involved is a sample statistic. In this case the sample statistic is the mean, and this is the reason why this particular sampling distribution is called the sampling distribution of the mean.

2. Sampling distribution of the mean. The sampling distribution of the mean is a normal distribution. The sampling distribution of $\bar{Y}$ will be well approximated by the normal distribution even if the distribution of Y itself is not normal. This result is an example of the force of the central limit theorem. In order to use the normal distribution to obtain probabilities, we need to know both its mean and its standard deviation. Once the mean and standard deviation are known, the sampling distribution of the mean can be used as described in Chapter 4 to answer simple questions about the probability of obtaining a sample mean within a designated range of values.

The mean of the sampling distribution of $\bar{Y}$ is μ. That is, the population mean of $\bar{Y}$ is the same as the population mean of Y. The standard deviation of the sampling distribution is symbolized $\sigma_{\bar{Y}}$. Its value is

$$\sigma_{\bar{Y}} = \frac{\sigma_Y}{\sqrt{n}}$$

The standard deviation of the sampling distribution of the mean is called the standard error of the mean. An important point is that, as the formula

$$\sigma_{\bar{Y}} = \frac{\sigma_Y}{\sqrt{n}}$$

indicates, the size of the standard error of the mean is determined by two factors: the standard deviation of Y (σ_Y) and the sample size (n).

3. Confidence intervals for μ (σ known). The first widely used application of the sampling distribution of the mean is to obtain a confidence interval that provides an interval estimate of the population mean, μ. A confidence interval is a range of values that claims to include μ with a specified probability such as .95. Such an interval would be referred to as a 95% confidence interval. Other confidence levels commonly used are 90% and 99%. Interval estimates have their counterpart in everyday language, as when someone states that "I am highly confident that my grade in this course will be between 70% and 80%."

The general form of the confidence interval of the mean is

$$CI = \bar{Y} \pm w.$$

When σ_Y is known, $w = z \times \sigma_{\bar{Y}}$. The value for z comes from tables of the normal curve and depends on the level of confidence desired. For the 95% confidence interval $z = 1.96$. The value of 1.96 is the z-score that leaves an area of .05 in the combined tails of the distribution. For a 99% confidence interval the value of z is 2.58.

4. Confidence intervals for μ (σ unknown). In practice, the value of σ_Y is typically unknown, and when this is the case the standard error must be estimated; the standard error $\sigma_{\bar{Y}}$ is replaced by $s_{\bar{Y}}$ where

$$s_{\bar{Y}} = \frac{s_Y}{\sqrt{n}}$$

As with all estimates, the estimated standard error is subject to sampling variability. That is, $s_{\bar{Y}}$ is a variable in the same way that $\bar{Y}$ is a variable. The consequence of this fact is that, unlike the usual ratio for the z-score which is

$$z = \frac{\bar{Y} - \mu}{\sigma_{\bar{Y}}},$$

the ratio

$$\frac{\bar{Y} - \mu}{s_{\bar{Y}}}$$

is *not* normally distributed. Instead of the normal distribution it has a t-distribution and the ratio

$$\frac{\bar{Y} - \mu}{s_{\bar{Y}}}$$

is referred to as a t-ratio. The difference between these two ratios is that the denominator of the z-score is a constant ($\sigma_{\bar{Y}}$) whereas the denominator of the t-ratio is a variable, ($s_{\bar{Y}}$), a difference that brings about the shift from a normal to a t-distribution.

The t-distribution has one parameter, the degrees of freedom, and the table of the distribution sets out values of t for critical probability values for different degrees of freedom. If the t-table does not include the exact value of the needed degrees of freedom ($df = 87$, for example) then use the closest value *below* the one needed. For large degrees of freedom the t-distribution is well approximated by the normal distribution.

When $\sigma_{\bar{Y}}$ must be estimated by $s_{\bar{Y}}$, the t-distribution, rather than the normal distribution, must be used to calculate confidence intervals. As before, CI $= \bar{Y} \pm w$, but now value of w is $w = t \times s_{\bar{Y}}$. The value of t is obtained from tables of the t-distribution and therefore depends on the degrees of freedom.

5. Meaning of confidence intervals. What does the expression "95% confidence" mean? Precisely how to interpret a confidence interval has been a controversial matter. The usual interpretation within the context of a long-run relative frequency interpretation of probability is that the confidence level (such as 95%) should be thought of as the long-run proportion of intervals that, over repeated sampling, would include the population mean, μ.

6. The width of a confidence interval. The width of a confidence interval is $2 \times w$, the difference between the upper and lower bound of the interval. The width of a confidence interval is a measure of the precision with which $\bar{Y}$ has been estimated. Narrower intervals reflect higher precision. Because achieving adequate precision is an important matter to experimenters, it is valuable to examine the determinants of a confidence interval's width. Several factors influence the width of a confidence interval. For a fixed level of confidence, the greater the variability of the data (measured by s_Y), the wider the interval. With everything else held constant, the larger the sample size, the narrower the interval. Note also that the greater the confidence demanded, the wider the interval must be; a 99% confidence interval will always be wider than a 95% confidence interval based on the same set of data.

7. Using the confidence interval to make decisions. Confidence intervals can also be used to make decisions about plausible values of μ. A simple definition of plausibility is that any value lying within the interval estimate of μ remains a plausible value. Values outside the interval can be declared implausible as possible values of μ. Note that plausibility and implausibility are relative concepts depending on the degree of confidence of the interval.

What You Should Know and Be Able to Do

■ Explain how sample statistics such as $\bar{Y}$ and s_Y can be regarded as variables and therefore also as random variables.

■ State the form of the sampling distribution of the mean and specify its parameters.

■ Given n, μ, and σ_Y, use the table of the normal distribution to specify the probability that $\bar{Y}$ will fall within a stated range of value.

■ Explain the difference between a point estimate and an interval estimate.

■ Explain what a 95% confidence interval is.

■ Given n, and σ_Y calculate a confidence interval for μ.

■ Given the values of n, $\overline{Y}$ and s_Y (σ_Y is unknown), calculate a confidence interval for μ.

■ Obtain the width of a confidence interval and explain why it provides a measure of precision.

■ Explain how the confidence interval can be used to make decisions about plausible values of μ.

Short-Answer Questions

1. A sample of 10 observations yields a sample mean of $\overline{Y} = 6.5$. Someone claims that this value of 6.5 can be thought of as constituting a sample of size one. Explain what this claim means, and then specify

   a. the population from which this "sample of size one" has been drawn.

   b. The form of the probability distribution from which it was drawn.

2. A second sample, this time of 15 observations, is drawn from the same population as in the previous question. This sample yields a sample mean of $\overline{Y} = 6.1$. Someone claims that the value of 6.1 is more likely to be closer to the population mean than is the value 6.5 obtained in the first sample. Do you agree? Justify your answer.

3. What is the relevance of the central limit theorem to the sampling distribution of the mean?

4. What is the difference between a point estimate and an interval estimate?

5. Two samples, one with $n = 15$ the other with $n = 40$, are drawn from a population with known variance but unknown mean. Each sample is used to calculate a 95% confidence interval for the unknown mean. Which interval would be wider? Justify your answer.

6. If calculated using the same data, which would be wider, a 95% or a 99% confidence interval?

7. Explain when you would use the normal distribution and when you would use the t-distribution in attempting to assign probabilities to values of $\overline{Y}$.

8. An investigator draws a sample of $n = 20$ and calculates the sample mean, $\overline{Y}$, the sample variance, s^2, and an estimate of the standard error of the mean, $s_{\overline{Y}}$. In calculating a 95% confidence interval the investigator mistakenly uses the normal distribution rather than the t-distribution. Without consulting the tables decide whether the resulting interval would be wider or narrower than it would have been had the correct t-distribution been used.

9. Two samples, each of $n = 45$, are drawn from different populations. The population variances are 15.0 and 45.0 for the first and second populations respectively, but the means are unknown. Each sample is used to calculate a 95% confidence interval for the unknown mean of the population from which it was drawn. Which interval would be wider? Justify your answer.

10. Suppose a theory predicts that the mean of a population is $\mu = 5.0$. In a test of the theory an investigator estimates μ by calculating a 95% confidence interval. The interval is 6.3 ± 1.1. Does this interval indicate that the predicted value of 5.0 is a plausible one?

Short Problems

Problem 1 A normal distribution has a mean, $\mu = 50.0$ and a variance, $\sigma^2 = 110.0$ What is the probability that the mean of a random sample of $n = 5$ observations will be greater than 48.0?

Problem 2 A normal distribution has a mean, $\mu = 60.0$ and a standard deviation, $\sigma = 9.0$. What is the probability that the mean of a random sample of $n = 12$ observations will be greater than 61.0?

Problem 3 Test scores are distributed with a mean of $\mu = 125.0$ and a standard deviation of $\sigma = 20.0$. What are the values that mark off the first and third quartiles of the sampling distribution of the mean for this population based on random samples of size $n = 10$ from the distribution of test scores?

Problem 4 Scores on an aptitude test are known to have a variance of $\sigma^2 = 440.0$ A random sample of 15 scores yields a sample mean of 75.0. What is the value of the standard error of the distribution the sample mean? Estimate the 95% confidence interval for the mean of the population from which this sample was drawn.

Problem 5 An aptitude test that has a mean $\mu = 50.0$ is administered to a sample of 200 students. The students are assigned randomly to one of two rooms, A and B. Room A seats 125 of these students, room B seats the remaining 75. The mean aptitude score is calculated separately for the sub-sample in each room. Which of the two sub-sample means is more likely to be greater than 52.0? Or are they equally likely? Give reasons for your answer.

Problem 6 A random sample of size $n = 24$ yields a sum of squares of 257.0. What is the estimated standard error of the mean of this distribution? What is the width of the 99% confidence interval for the mean of the population from which this sample was drawn?

Problem 7 An experimenter calculates the mean, variance, and the 95% confidence interval for the mean for data from a perception experiment with $n = 45$ subjects. However, closer examination of the raw data shows that whereas 42 of these participants obtained scores ranging from 53% to 89% correct responses, 3 participants scored only 2%, 4% and 0% correct. The experimenter decides to remove these outliers and recalculate various statistics with the reduced sample of $n = 42$. What would the effect of discarding these residuals be on the width of the 95% confidence interval?

Problem 8 Which (if any) of the following changes would (other things being held constant) increase the width of a confidence interval estimate of the mean based on sample data?

    a. Increasing the sample size.

    b. Mistakenly using the normal rather than the t-distribution.

    c. Changing the confidence level from 99% to 95%.

    d. Reducing the variance of the scores (e.g. through better experimental control).

Problem 9 A simple random sample of $n = 25$ is drawn from a distribution of IQ scores that are normally distributed with a mean of 100 and a standard deviation of 15.

    a. What is the probability that the mean of this sample will be greater than 103?

    b. Another random sample of $n = 25$ is drawn from this distribution. What is the probability that the mean of this sample will fall in the interval 95 to 105?

Problem 10 A random sample of reaction times ($n = 23$) yields a 99% confidence interval for the mean of 872 – 978 milliseconds.

    a. What was the estimated standard error of the mean of the distribution of these 23 reaction times?

    b. What was the estimated standard deviation of the distribution of these 23 reaction times?

    c. What is the 95% confidence interval for the population mean reaction time?

Exercises in the Analysis of Realistic Data

Exercise 1. Authoritarianism

Exercise 7 of Chapter 2 described an investigation of the relationship between the authoritarianism of parents and various characteristics of their children. When tested on a random sample of 150 mothers, the scale produced scores with a mean of 45.7 and a standard deviation of 9.1. Calculate a 95% confidence interval for the mean of the population from which this sample was drawn. The researcher drew a second sample of mothers ($n = 65$, Data Set CHILD_3.dat.) and obtained a mean of 48.8 and a standard deviation of 9.67. Use these results to calculate a second 95% confidence interval for the mean of the population from which this sample was drawn. Discuss the reasons for the differences between these two confidence intervals.

Exercise 2. Developing an aptitude test (Data Set TEST_1.dat)

Imagine once again that you are the Director of Psychological Services charged with the responsibility of constructing and evaluating a scholastic aptitude test. In Exercise 9 of Chapter 2 the mean and standard deviation of the sample ($n = 200$) was found to be 86.36 and 28.43 respectively. Using these estimates of μ and σ, obtain an estimate of the standard error of the mean. Then construct a 95% confidence interval for μ.

In a follow up study, a sample of 100 participants were given the scholastic aptitude test. The mean and standard deviation of this sample was found to be 88.06 and 27.40, respectively. Using these estimates, obtain an estimate of the standard error of the mean. Then use this estimate to construct a second 95% confidence interval for μ. Why is this second confidence interval wider than the first?

Exercise 3. An extroversion scale (Data Set TEST_2.dat)

In Exercise 10 of Chapter 2 the Extroversion scale was administered to a sample $n = 120$ participants. The mean and standard deviation were found to be 58.97 and 14.22 respectively. Using these values as estimates of μ and σ, obtain an estimate of the standard error of the mean. Then construct a 99% confidence interval for μ.

In a follow up study, a second sample of 120 participants were given the extroversion scale. The mean and standard deviation of this sample was found to be 56.56 and 17.49 respectively. Using these estimates, obtain an estimate of the standard error of the mean. Then use this estimate to construct a second 99% confidence interval for μ. Why is this second confidence interval wider than the first?

6 EXPERIMENTS WITH TWO INDEPENDENT GROUPS

Overview

Chapter 6 describes the analysis of two-condition experiments using two independent groups. The goal of the analysis is to evaluate the true difference between the means of the conditions and to decide which of two models (the null or the full model) are appropriate. Two approaches to these tasks are presented:

- Obtaining confidence intervals for $\mu_1 - \mu_2$.

- Testing directly the null hypothesis that $\mu_1 = \mu_2$.

1. Confidence intervals for $\mu_1 - \mu_2$. The method for calculating a confidence interval for the difference between two means is directly analogous to that used for a single mean. For the confidence interval of $\mu_1 - \mu_2$:

- The point estimate that forms the center of the confidence interval is $\overline{Y}_1 - \overline{Y}_2$ rather than $\overline{Y}$.

- The standard error is $s_{\overline{Y}_1 - \overline{Y}_2}$ rather than $s_{\overline{Y}}$.

Because $s_{\overline{Y}_1 - \overline{Y}_2}$ is a sample-based estimate and not a parameter, the relevant sampling distribution is the t-distribution rather than the normal distribution. With a single mean the degrees of freedom were $(n - 1)$. With two means the degrees of freedom are the sum of the degrees of freedom for each group: $(n_1 - 1) + (n_2 - 1) = (n_1 + n_2 - 2) = (N - 2)$. If the two groups are of equal size, then the degrees of freedom can be written as $2(n - 1)$. Once $\overline{Y}_1 - \overline{Y}_2$ and $s_{\overline{Y}_1 - \overline{Y}_2}$ have been calculated, and the appropriate value of the t-ratio has been obtained from the table of the t-distribution, then the confidence interval is CI $= |\overline{Y}_1 - \overline{Y}_2| \pm w$, where $w = t \times s_{\overline{Y}_1 - \overline{Y}_2}$.

2. Using the confidence interval to make decisions. This confidence interval can be used to decide whether or not a specified value of $\mu_1 - \mu_2$ is plausible. Usually the specific value of interest is zero—the null hypothesis, H_0. If the confidence interval contains the value of zero within its range, then H_0 remains plausible. If it does not, then the null hypothesis that $\mu_1 - \mu_2 = 0$ can be rejected as implausible. Remember that the interval will include zero if the observed difference is less than w. That is, if $|\overline{Y}_1 - \overline{Y}_2| < w$ then retain H_0.

The criterion that distinguishes plausible from implausible is set by the interval's confidence level. Suppose for example that the confidence level is 95% and the interval does not include zero. Then the null hypothesis is implausible in the following sense: If the difference is really zero, then intervals that do not include zero are unlikely to arise—they occur with a probability of less than .05.

Rejecting the null hypothesis amounts to rejecting the null model in favor of the full model. This full model requires a distinct parameter for each condition (μ_1 and μ_2). If the evidence does not support the claim that the means are different, then the conditions are assumed to make no difference and so subscripts can be dropped and the more parsimonious null model ($Y = \mu + e$) is accepted as an adequate account of the data for both groups.

3. Direct hypothesis testing. It is possible to evaluate the null hypothesis and the associated null model directly by evaluating the probability associated with the observed difference $\bar{Y}_1 - \bar{Y}_2$. The rationale for this evaluation is the same as that used with the confidence interval. If a difference as large as that observed is improbable when in fact the true difference is zero, then reject the null hypothesis. Put simply, large differences rarely arise from a situation in which the true difference is zero; so if a large difference *is* observed then the claim of a zero difference should be discarded. A physician might use the same logic. A temperature of 102 degrees is rarely observed in a healthy person. If such a large difference from normal temperature is observed, then the physician rejects the hypothesis that the patient is in good health.

4. The *t*-distribution. The probability associated with an observed difference between sample means is obtained using the *t*-distribution. Note that, as with the confidence interval, the *t*-distribution is the relevant distribution because the population variances are unknown, and thus the standard error must be estimated from the data. If the null hypothesis is true then the probability associated with the observed difference is obtained by calculating the *t*-ratio

$$t = \frac{(\bar{Y}_1 - \bar{Y}_2)}{s_{\bar{Y}_1 - \bar{Y}_2}}$$

The degrees of freedom are, of course, the same as those used for the confidence interval because the *t*-ratio is based on exactly the same standard error. The probability associated with this value of the *t*-ratio can be obtained by consulting the table of the *t*-distribution. Locating the row in the table for the appropriate degrees of freedom, the probability can be located relative to the three probability levels of .1, .05, and .01 found in the table.

The customary minimum criterion for rejecting the null hypothesis is a probability of .05. Using this criterion is mathematically equivalent to basing the decision on the 95% confidence interval as described above. If a criterion of .01 were used, then the decision rule is mathematically equivalent to basing the decision on the 99% confidence interval.

5. Significance levels. If the probability value obtained is less than the criterion (less than .05, for example), then the difference is described as "statistically significant." Some researchers prefer to report the level of significance by reporting the probability associated with the *t*-ratio. For example, if the table shows the *t*-ratio to have a probability of less than .01 then they would describe the difference as "significant at the .01 level." This method of reporting the significance level was described as the Fisher tradition. The basic strategy of this tradition is to find the probability associated with the *t*-ratio, and then decide whether or not this probability is low enough to reject the null hypothesis. Usually, "sufficiently low" means less than .05, but it is not uncommon to find probabilities of .06 or even .1 described as "marginally significant."

6. Decision error rates—Type-1. Neyman and Pearson formulated a position somewhat different from that of Fisher. In this tradition two competing hypotheses are formulated: The null hypothesis: $H_0: \mu_1 - \mu_2 = 0$ and an alternative hypothesis, most commonly $H_1: \mu_1 - \mu_2 \neq 0$. Neyman and Pearson re-interpreted the significance level as the Type-1 error rate (denoted α) and insisted that the investigator set its value prior to any data analysis. The value set for α is an error rate in the following sense: It is a probability associated with a decision error—that of rejecting a null hypothesis that is actually true. You can think of a Type-1 error as a false alarm—claiming a difference when none really exists.

7. Decision error rates—Type-2. Neyman and Pearson also defined a Type-2 error, and the probability associated with this error is denoted by β. A Type-2 error is a failure to detect a real difference. Think of it as a miss—a failure to detect a signal that is really present. Type-2 error rates are influenced by two factors: The size of the true difference and the size of the standard error $s_{\bar{Y}_1 - \bar{Y}_2}$. In common sense terms these two factors state that (a) the larger the differences the more easily it can be detected and (b) the noisier the data (the larger is $s_{\bar{Y}_1 - \bar{Y}_2}$) the more difficult it is to detect a difference of any given size.

Attention to Type-2 errors can help resolve the problem of the marginally significant difference. It does this by drawing attention to the importance of the magnitude of the difference between the means, rather than simply whether or not an observed difference is significant. If an analysis leads to the retention of the null hypothesis, then it is important to know something about the power of the statistical test.

8. The power of a statistical test. The power of a test refers to the probability of detecting a difference of a specified size. Numerically, its value is $1 - \beta$. Statistical power is analogous to the power of a microscope: The greater the power of a microscope, the smaller the organism that it can reliably detect. The power of a microscope might therefore be described as the probability that it will detect an organism of a given size. If a microscope fails to detect a suspected organism, then it is reasonable to ask whether it was powerful enough to have a high probability of detecting an organism of such size. When the null hypothesis is retained it is reasonable to ask whether the test had sufficient power to detect a difference of a specified magnitude.

What You Should Know and Be Able to Do

■ From the description of an experiment, decide whether or not it is an independent-groups design. An independent-groups design may be either a completely randomized design or the groups may be based on a natural predictor variable. In a completely randomized design, participants are assigned randomly to the conditions. With a natural predictor variable (such as age or sex), random assignment is impossible, but the groups will be independent provided the selection of a participant for one group has no influence on whether or not a particular participant will, chosen for the other group.

■ Calculate the estimated standard error of the difference between two means. For groups of equal size (n participants in each) the formula is

$$s_{\bar{Y}_1 - \bar{Y}_2} = \sqrt{\frac{2MS_e}{n}}$$

■ Calculate the degrees of freedom for independent-groups design with two groups or conditions. The degrees of freedom are $2(n-1)$ for groups of equal size. For unequal groups the degrees of freedom are $(n_1 - 1) + (n_2 - 1) = (n_1 + n_2 - 2) = (N-2)$.

■ Know why the t-distribution is the appropriate sampling distribution and how to use the table of the t-distribution to obtain the value of the t-ratio corresponding to the desired level of confidence.

■ Calculate the semi-width of the confidence interval: $w = t \times s_{\bar{Y}_1 - \bar{Y}_1}$.

■ Calculate the confidence interval as $|\bar{Y}_1 - \bar{Y}_2| \pm w$.

■ Use the confidence interval to decide whether or not the null hypothesis should be rejected and thus whether the data justify the full model as opposed to the more parsimonious null model.

■ Calculate a t-ratio directly and use it to evaluate the null hypothesis of zero difference. The t-ratio is the difference between the sample means divided by the estimated standard error of the difference. The difference between two means is declared statistically significant if the t-ratio is sufficiently large to be improbable under circumstances in which the null hypothesis is true. The criterion of improbability is called the significance level and is commonly set at a probability of .05.

■ Understand the difference between a Type-1 and a Type-2 error. A Type-1 error is a false alarm—mistakenly declaring a difference to exist. A Type-2 error is a miss—a failure to detect a genuine difference. The probability of a Type-1 error is designated α, a Type-2 error β.

■ Understand the trade-off relation between Type-1 and Type-2 errors. Other things being constant, reducing the Type-1 error rate (α) increases the probability of a Type-2 error (β).

■ Understand the ambiguity that exists when the null hypothesis is retained. Either the true difference is zero, or there is a non-zero difference and a Type-2 error has been made. In the case of a Type-2 error, the test lacked sufficient power to detect a difference of specified size.

■ Know that (and why) one way of increasing the precision (power) of a statistical test is by increasing the sample size.

Short-Answer Questions

1. Decide whether each of the following statements is true or false and give your reason.

   a. Holding other factors constant, the larger the value of $s_{\bar{Y}_1 - \bar{Y}_2}$ the wider will be the 95% confidence interval.

   b. Holding other factors constant, the larger the sample size the wider will be the 95% confidence interval.

   c. Holding other factors constant, the 95% confidence interval will always be wider than the 99% confidence interval.

d. In designing an experiment, an investigator aims for as wide a confidence interval as possible because the wider the confidence interval, the more likely it is to include the true difference.

e. Using a 95% confidence interval and a t-ratio with a significance level of .05 will always lead to the same decision as to whether the null hypothesis should be rejected or accepted.

f. The half-width of a 95% confidence interval (w) is the minimum difference needed to reject the null hypothesis at the .05 significance level.

g. An experiment using an independent-groups design has 12 participants in each condition. The degrees of freedom for the t-ratio for this experiment are 23.

h. Random assignment of participants to conditions ensures that the groups are matched on factors such as individual differences in ability and personality.

i. The null model is a more parsimonious account of the data than the full model.

j. If the residuals are independently and normally distributed with equal variances, then the normal distribution rather than the t-distribution can be used to calculate a confidence interval for the difference between the means.

k. An investigator calculates a 99% confidence interval, finds that it does not include the value of zero and so rejects the null hypothesis. Had the investigator used these data to calculate a 95% confidence interval instead, we can be certain that it, too, would not have included zero.

l. Increasing the precision of an experiment by increasing the sample size would increase the likelihood of committing a Type-1 error.

m. Holding everything else constant, decreasing the Type-1 error rate increases the likelihood of committing a Type-2 error.

n. If the null hypothesis is false, you can never commit a Type-2 error, no matter how small the sample.

o. In testing hypotheses about differences between two means, an experimenter mistakenly uses 36 instead of 20 degrees of freedom when looking up the tables of the t-distribution to find the critical value. Assuming all other calculations are correct, the effect of this error will be to increase the probability of a Type-1 error.

p. An experimenter sets $\alpha = .05$ as the significance level, obtains a t-statistic less than the critical value and therefore does not reject the null hypothesis. This statement means that the probability of the decision being erroneous is .05.

q. The 95% confidence interval for the difference between two means was found to be 1.2 ± 0.9. This result implies that (using a Type-1 error rate of $\alpha = .05$) we can reject the hypothesis that the true difference is zero.

r. In testing hypotheses about differences between two means, using the normal distribution as an approximation to the t-distribution will inflate the likelihood of a Type-1 error.

s. The power of a statistical test can be described as the probability of not committing a Type-2 error.

t. In planning an experiment, an investigator first decides to use a Type-1 error rate of $\alpha = .01$ but finally plans on using $\alpha = .05$. The effect of this change of plan would be to reduce the likelihood of a Type-2 error.

2. Read the brief descriptions of the following three studies. One of these studies uses a completely randomized design, one is a quasi-experiment with an independent groups design, and one is not an independent-groups design at all. Classify each study.

a. In a study of the effects of birth order on scholastic achievement, a random sample of 50 two-child families was drawn. The grade-point average was obtained for the both the first- and second-born child from each of the 50 families.

b. In a comparison of different teaching methods on math performance, a random sample of 50 grade-nine children was selected from each of two schools that had used the different methods. The mean math grades for the two groups were compared.

c. In a comparison of two different study methods on second-language learning, a sample of 100 children was selected and randomly divided into two groups of 50 each. Each group used one of the two study methods. Mean language achievement scores for the two groups were compared.

3. An experiment has two conditions and uses an independent-groups design. An investigator has obtained an estimate of 3.0 for the standard error of the difference between the two means. The estimate is based on 60 degrees of freedom. Using a significance level (Type-1 error rate) of .05, the investigator claims that any observed difference between the means of 6.0 or more is incompatible with the null hypothesis of zero difference. Explain how the value of 6.0 was obtained, and the reasoning behind the decision rule to reject the null hypothesis for any observed difference greater than 6.0.

4. In an effort to explain the difference between a Type-1 and a Type-2 error, an instructor draws the following analogy.

A mining engineer obtains a sample of rock, finds no gold, and concludes that there is no gold in the area. The company's shares drop in price. However, it turns out that there really is gold in the area and that the engineer just happened to drill in a spot containing no gold. A second engineer working for a different company in a different area obtains a sample of rock and finds traces of gold. He declares "strike" and the company's shares double in price. However, it turns out that there is no gold in the general area and the engineer's sample was a fluke.

Which engineer made the equivalent of a Type-1 error and which a Type-2 error? Explain your answer.

Short Problems

Problem 1 An experiment uses 28 participants, randomly assigning 14 to each of two conditions. What is the tabled value of the t-statistic you would use to calculate the 95% confidence interval for the true difference between the means of these conditions?

Problem 2 An experiment uses 36 subjects, randomly assigning 18 to each of two conditions. What is the tabled value of the t-statistic you would use to calculate the 99% confidence interval for the true difference between the means of these conditions?

Problem 3 An experiment has two conditions. Participants are assigned randomly and independently to the two conditions such that there are 20 in each condition. The difference between the two resulting sample means is 8.70. The experimenter calculates the appropriate t-ratio and obtains a value of $t = 2.30$.

    a. What was the estimated standard error of the difference between the means?

    b. A reviewer of the paper reporting this result requests that the experimenter also give a 99% confidence interval for the true difference between the means. Calculate this confidence interval.

    c. With a Type-1 error rate of .01, should you reject the null hypothesis that the true difference is zero?

    d. Write out the model you judge to be appropriate for the data from this experiment.

Problem 4 An experiment using an independent groups design is analyzed by calculating a t-ratio. There are 21 randomly assigned participants in each of the two conditions, and the standard error of the difference between the means is found to be 4.4. The Type-1 error rate is set at $\alpha = .05$. What is the minimum absolute difference between the means that would have to be obtained in order to reject the model under the null hypothesis: $Y_1 = Y_2 = \mu + e$ in favor of the full model $Y_1 = \mu_1 + e$; $Y_2 = \mu_2 + e$?

Problem 5 A drug experiment has two conditions, experimental and control. The experimental condition is designated 1 and the control condition is designated 2. Scores for the experimental condition are therefore designated Y_1, and scores in the control condition Y_2. The results yield the following values:

Condition	n	Mean	Sum of squares
Experimental	30	$\bar{Y}_1 = 10.0$	$SS_1 = 45.0$
Control	30	$\bar{Y}_2 = 6.0$	$SS_2 = 55.0$

The experimenter fits two models, Models 1 and 2.

Model 1: $Y_1 = \mu_1 + e.$

                   $Y_2 = \mu_2 + e.$

Model 2: $Y_1 = Y_2 = \mu + e$

    a. What is the estimated residual component (e) for each of these models for:

        (i) a score of $Y_1 = 7.0$ from the experimental condition?

        (ii) a score of $Y_2 = 7.0$ from the control condition?

b. What is the value of the sum of squares of the residuals for Model 1?

c. What is the value of MS_e?

d. What is the estimated effect size (Cohen's **d**) for these data?

Problem 6 In an experiment designed to study the effects of fatigue on skilled performance, 30 subjects are assigned randomly and independently to one of two conditions (15 subjects in each). The response measure is the number of errors made over a 30-minute test period. The mean error score for the high-fatigue condition was 45.6 and for the non-fatigued control group it was 32.8. The value of MS_e was 723.0.

a. Calculate the value of the standard error you would use to calculate a confidence interval for the true difference between the means of the two conditions.

b. What is the critical tabled value of the t-statistic you would use to calculate a 95% confidence interval for the difference between the means of the two conditions?

c. Using a Type-1 error rate of $\alpha = .05$, what is the minimum difference between the means of the two conditions needed to reject the null hypothesis of zero difference?

d. Using the obtained means for each condition, calculate the t-ratio you might use to evaluate the null hypothesis. Convince yourself that using the confidence interval, the minimum difference, or the t-ratio must all lead to identical decisions.

e. Write out the model you judge to be appropriate for the data from this experiment.

Exercises in the Analysis of Realistic Data

Set 1: Understanding Memory

Exercise 1. The generation effect (Data Set MEM_02.dat)

The data from this experiment were examined in Exercise 2 of Chapter 2. A total of 50 participants took part in this experiment, 25 being randomly assigned to each of two conditions. Participants in the "read" condition saw a list of 40 common words such as **small** and **happy** and were instructed to read them out loud in preparation for a subsequent recall test. Participants in the "generate" condition saw a list of words along with the first letter of a second word. They were instructed to generate a word that started with this letter and meant the opposite of that word. For example, if they saw **large s___**, they responded "small" or if they saw **sad h____** they responded "happy." In this way both groups spoke an identical set of words, the "read" groups by reading them directly, the "generate" group by recalling the target word in response to the cue word.

Following the presentation of the words, all participants were asked to recall many of the spoken words as they could remember. The results of basic calculations were as follows:

Condition	n	Mean	SS_Y	SS_{total} *
Read	25	9.52	134.2	649.0
Generate	25	14.52	202.2	

a. Calculate the value of MS_e.

b. Describe the effect size in this experiment by calculating an estimate of Cohen's **d**.

c. Calculate a 95% confidence interval for the difference between the means of the two conditions.

d. Calculate R^2, the percentage reduction in the sum of squared residuals achieved by the full model compared to the null model.

e. Do the data justify the full model, or is the simpler null model an adequate account?

f. What conclusion would you draw about the effect of the experimental manipulation?

Exercise 2. Memory for chess positions I (Data Set MEM_03.dat)

The data from this experiment were examined in Exercise 3 of Chapter 2. Recall that this study compared the memory of chess experts for chess positions with the memory of novices for the same positions. Two groups were selected. Group A consisted of 22 chess experts, Group B of 22 chess novices. The response measure was the number of board positions correctly recalled. The results of basic computations were as follows:

Group	n	Mean	SS	SS_{total}
Experts	22	11.86	388.6	768.73
Novices	22	7.59	179.3	

a. Calculate the value of MS_e.

b. Describe the effect size in this experiment by calculating an estimate of Cohen's **d**.

c. Calculate a 99% confidence interval for the difference between the means of the two conditions.

d. Calculate R^2, the percentage reduction in the sum of squared residuals achieved by the full model compared to the null model.

e. Using the 99% confidence interval, decide if the data justify the full model or whether the simpler null model would provide an adequate account?

---

* Note that in this and the following exercises, SS_{total} is the sum of squared residuals under the null model. Recall that SS_{model} is $SS_{total} - SS_e$.

f. What conclusion would you draw about the effect of the difference between novices and experts in memory for chess positions?

Exercise 3. Memory for chess positions II (Data Set MEM_04.dat)

A possible interpretation of the chess experiment in Exercise 2, is that chess experts have naturally better memories to start with, and that rather than being the *result* of their chess experience, their superior memories are a major cause of their becoming expert.

To explore this possibility, a second study was conducted, again comparing the memory of expert and novice chess players. There were 28 participants in each group. In the original study the board positions to be remembered were meaningful game positions. In this second study the chess pieces were arranged in random positions on the board. As in the first study, the response measure was the number of board positions correctly recalled. The data from this experiment were examined in Exercise 3 of Chapter 2. The results of basic computations were as follows.

	n	Mean	SS	SS_{total}
Experts	28	7.75	131.25	300.55
Novices	28	6.93	159.86	

a. Calculate the value of MS_e.

b. Describe the effect size in this experiment by calculating an estimate of Cohen's **d**.

c. Calculate a 95% confidence interval for the difference between the means of the two conditions.

d. Calculate R^2, the percentage reduction in the sum of squared residuals achieved by the full model compared to the null model.

e. Calculate a t-ratio and, with a Type-1 error rate of $\alpha = .05$, use it to decide whether the data justify the full model or whether the simpler null model would provide an adequate account. Check that this decision is consistent with the one you might have made on the basis of the confidence interval.

f. What conclusion would you draw about the effect of the difference between novices and experts in memory for chess positions?

Exercise 4. Chess expertise and children's memory (Data Set MEM_05.dat)

The following experiment is based on a study reported by Chi (1978). A well-known result in developmental psychology is that a 10-year old child's short-term memory capacity is less than that of an adult. If asked to repeat a string of digits an average 10-year old can repeat between 5 and 6 digits whereas a typical adult can repeat 7 or 8. Can suitable training reverse this difference? Consider 10-year-olds who are chess experts. These children have spent hundreds of hours studying chess positions and it is interesting to ask whether this extensive experience has led to a short-term memory for chess positions that is superior to that of an adult novice chess player. To answer this question 10-year-old chess experts were compared with adult novice chess players on their memory for the positions of chess pieces. Samples of 15 adults and 15 10-year-olds were used. The response measure is the number of chess positions correctly recalled. Hypothetical data from this study are given below.

Data Set MEM_05.dat

Children

7   6   13   10   5   4   6   10   8   10   10   7   9   12   10

Adults

3   3   8   7   7   9   6   6   5   9   5   2   5   3   6

	n	Mean	SS	SS_{total}
Children	15	8.47	93.73	222.97
Adults	15	5.60	67.61	

a.   Calculate the value of MS_e.

b.   Describe the effect size in this experiment by calculating an estimate of Cohen's **d**.

c.   Calculate a 95% confidence interval for the difference between the means of the two conditions.

d.   Calculate R^2, the percentage reduction in the sum of squared residuals achieved by the full model compared to the null model.

e.   Calculate a t-ratio and, with a Type-1 error rate of $\alpha = .05$, use it to decide whether the data justify the full model or whether the simpler null model would provide an adequate account. Check that this decision is consistent with the one you might have made on the basis of the confidence interval.

f.   What conclusion would you draw about the effects of training on short-term memory?

Set 2: Child Behavior

Exercise 5. Familiarity and compliance (Data Set CHILD_1.dat)

The data from this experiment were examined in Exercise 5 of Chapter 2. Are children more likely to comply promptly to a request made by a parent or by a stranger? Recall that in this experiment, seven-year-old children were taken to a laboratory playroom that has intentionally been left messy. After entering the room the child was asked to clean it up. In one condition the child's parent made the request. In a second condition the request was made by a stranger (the investigator).

A total of 50 seven-year-old children participated in the experiment. They were assigned randomly to the conditions, 25 in each. The response measure was the time taken (latency) for the child to begin the task.

The investigator in this study would like to decide whether there is any difference between the two conditions. The mean for the parent-request condition was 30.0 and for the stranger-request condition the mean was 19.9. The sum of squares for the parent-request condition was 2297.0, and for the stranger-request condition it was 1534.6. The value of SS_{total} was found to be 5101.7.

   a.   Calculate the value of MS_e.

   b.   Describe the effect size in this experiment by calculating an estimate of Cohen's **d**.

   c.   Calculate a 95% confidence interval for the difference between the means of the two conditions.

   d.   Calculate R^2, the percentage reduction in the sum of squared residuals achieved by the full model compared to the null model.

   e.   Calculate a t-ratio and, with a Type-1 error rate of $\alpha = .05$, use it to decide whether the data justify the full model or whether the simpler null model would provide an adequate account. Check that this decision is consistent with the one you might have made on the basis of the confidence interval.

   f.   What conclusion would you draw about the difference between parent and stranger instructions?

Exercise 6. Mood and compliance (Data Set CHILD_5.dat)

Are children more likely to comply promptly to a request if they are in a good mood? In this experiment, as in the previous one, seven-year-old children were taken to a laboratory playroom that has intentionally been left messy. In one condition the child was given a small gift and told that they could take it home with them. The child was then asked to clean up the room. Children in a control group received no gift but were simply asked to clean up the room.

    A total of 46 seven-year-old children participated in the experiment. They were assigned randomly to the conditions, 23 in each. The response measure was the time taken (latency) for the child to begin the task. The data were as follows.

Data Set CHILD_5.dat

Control (neutral mood) condition

```
21  24   18   23   21   29   27   16   24   17   22   24
17  22   27   18   24   19   14   20   25   15   20
```

Gift (good mood) condition

```
5   19   12   13   19   10   21   25   24   23   23   26
14   8   11   15   16   20   12   20   6    16   17
```

1. Construct a stemplot for each condition and check the data for anomalies.

2. The investigator in this study would like to decide whether there is any difference between the two conditions. The mean for the neutral mood control condition was 21.2 sec. For the gift (good mood) condition the mean was 16.3 sec. The sum of squares for the neutral control condition was 359.3 and for the gift condition it was 808.9. The value of SS_{total} was found to be 1440.9.

   a.   Calculate the value of MS_e.

   b.   Describe the effect size in this experiment by calculating an estimate of Cohen's **d**.

   c.   Calculate a 95% confidence interval for the difference between the means of the two conditions.

d. Calculate R^2, the percentage reduction in the sum of squared residuals achieved by the full model compared to the null model.

e. Calculate a t-ratio and, with a Type-1 error rate of $\alpha = .05$, use it to decide whether the data justify the full model or whether the simpler null model would provide an adequate account. Check that this decision is consistent with the one you might have made on the basis of the confidence interval.

f. What conclusion would you draw about the effect of mood on compliance?

Exercise 7. Attention and compliance (Data Set CHILD_6.dat)

Are children likely to comply to a request less promptly if they are currently engaged in an activity as compared to the same request made when the child is not involved in an activity?

Active

28 20   25   33   28   42   38   23   35
24 29   10   24   30   39   26   34   27

Nonactive

16 24   17   18   24   11   25   18   19
25 10   27   31   30   29   29   32   20

The investigator in this study would like to decide whether there is any difference between the two conditions. The mean for the active condition was 28.6 sec. For the nonactive condition the mean was 22.5 sec. The sum of squares for the active condition was 984.28 and for the nonactive condition it was 760.50. The value of SS_{total} was found to be 2080.89.

a. Calculate the value of MS_e.

b. Describe the effect size in this experiment by calculating an estimate of Cohen's **d**.

c. Calculate a 95% confidence interval for the difference between the means of the two conditions.

d. Calculate R^2, the percentage reduction in the sum of squared residuals achieved by the full model compared to the null model.

e. Calculate a t-ratio and, with a Type-1 error rate of $\alpha = .05$, use it to decide whether the data justify the full model or whether the simpler null model would provide an adequate account. Check that this decision is consistent with the one you might have made on the basis of the confidence interval.

f. What conclusion would you draw about the effect of attention on compliance?

7 LARGER EXPERIMENTS WITH INDEPENDENT GROUPS: ANALYSIS OF VARIANCE

Overview

1. Models for experiments with more than two conditions. Chapter 7 describes the analysis of experiments using an independent groups design with more than two groups. With more than two conditions, the full model is a straightforward extension of the two-condition case. It is convenient to use subscripts rather than to list all the conditions, so the model can be written as $Y_i = \mu_i + e$.

The sum of squared residuals for the full model is also a straightforward extension of the two-condition case. It is formed by adding together the sum of squared residuals for each of the k conditions. The value of MS_e is obtained in the usual way, by dividing the sum of squared residuals by the degrees of freedom. There are $n - 1$ degrees of freedom for each of the k conditions; that is, MS_e is based on $k(n - 1)$ degrees of freedom in total.

Two important new conceptual issues arise with these larger experiments, and they both stem from the fact that with more than two conditions, the overall impact of the predictor variable can no longer be captured by a single comparison between two means. Larger experiments generate multiple comparisons between the means of pairs of conditions. An experiment with four conditions, for example, has six possible comparisons.

2. Simultaneous confidence intervals. The first new conceptual issue demanded by multiple comparisons is exemplified in the distinction between simultaneous and individual confidence intervals. With individual confidence intervals, the confidence level refers to each interval considered separately. However, if this confidence level is .95, then the probability that *all* of the intervals will include the true difference is less than .95. The larger the number of comparisons, the more chances there are that at least one of the intervals will not include the true difference; just as the more dice you roll, the greater is the chance that at least one of them will land on six. For this reason most investigators prefer to calculate simultaneous confidence intervals. One method of calculating simultaneous intervals, called the Tukey method, replaces the t statistic used for individual confidence intervals with the q statistic obtained from the Studentized range distribution. The Studentized range distribution has two parameters: the number of conditions, k, and the degrees of freedom, $k(n - 1)$.

3. Error rates. Hypotheses about comparisons can be tested directly. In doing so, a distinction is drawn between error rates per comparison and error rates per experiment. This distinction parallels that between simultaneous and individual confidence intervals. A distinction is also drawn between planned and post hoc methods of testing comparisons. A method of planned comparison, such as the Bonferroni method, designates a small number of comparisons prior to the experiment. A method of post comparison such as Tukey's HSD (honestly significant difference) permits the testing of all comparisons. Methods of planned comparison have the advantage of a smaller Type-2 error rate than post-hoc comparisons.

The second new conceptual issue arises from the need to have an overall test of the null versus the full model. With just two conditions, this test could be accomplished using the t-ratio. With more than two conditions, the method of analysis is known as analysis of variance.

4. Analysis of variance. Analysis of variance introduces some new terminology for some familiar concepts. The difference in the sum of squared residuals between the null and the full model provides a measure of the overall differences between the condition means. It is termed the between-condition sum of squares, and written $SS_{between}$. The sum of squared residuals for the full model, is called the within-condition sum of squares, and written SS_{within}. The sum of squared residuals for the null model, is called the total sum of squares, and written SS_{total}. The important equation is that

$$SS_{total} = SS_{between} + SS_{within}$$

This equation states that the total sum of squares can be partitioned into two components: one that reflects differences among the condition means, and one that measures noise.

The sum of squares between conditions, and the sum of squares within conditions, can be converted to mean squares by dividing the sum of squares by their respective degrees of freedom. For the mean square within conditions (MS_{within}), the degrees of freedom are $k(n - 1)$; for the mean square between conditions ($MS_{between}$), the degrees of freedom are $k - 1$. Note that MS_{within} is the same as MS_e. If we add the degrees of freedom for $SS_{between}$ with those for SS_{within} we get the degrees of freedom for SS_{total}.

5. Rewriting the model. The full model can be rewritten by expressing each condition mean as a deviation from the grand mean. This rewriting led to the model being expressed as:

$$Y_i = \mu + \alpha_i + e$$

where α_i is the difference between μ_i and the grand mean, μ. The estimate of α_i is a_i and is equal to $\overline{Y}_i - \overline{Y}$. The values of a_i are related to $SS_{between}$ in a very simple way:

$$SS_{between} = n \times \sum_i a_i^2$$

6. *F*-distribution. The null hypothesis, and thus the null model, can be evaluated using the ratio

$$F = \frac{MS_{between}}{MS_{within}}$$

This ratio has a probability distribution known as the *F*-distribution which has two parameters: the degrees of freedom for the numerator mean square, and the degrees of freedom for the denominator mean square. If the F-ratio leads to a rejection of the null hypothesis, then Fisher's LSD (least significant difference) procedure can be used to evaluate comparisons. This procedure is less conservative than is the Tukey HSD method. When analysis of variance procedures are applied to an independent-groups design with just two conditions, the resulting F-ratio equals the square of the corresponding t-ratio.

7. Factorial designs. Larger experiments are often designed with a factorial layout consisting of two (or more) factors with each factor having two or more levels. A factor is a set of conceptually related conditions (such as different drugs or different methods of instruction), and a level refers to each condition within a factor. In a two-factor experiment, the factors can be labeled A and B. If the A factor has a levels and the B factor has b levels, then the experiment is described as an $a \times b$ factorial design. An experiment that evaluated three different forms of psychotherapeutic treatment of depression, each combined with either a placebo or an antidepressant drug (six conditions in all) would be described as a two-way factorial experiment with three levels of the first factor and two levels of the second factor or, briefly, a 3×2 factorial design.

Models of factorial designs distinguish between main effects and their interactions. In a two-factor experiment, the term "main effects" refers to differences among the means of the levels of one factor averaged over the levels of the other factor. An interaction reflects the non-additivity of the main effects—a situation in which the influence of one factor is different at different levels of the other factor.

The analysis of variance summary table for a factorial design records the partitioning of the $SS_{between}$ term into its three contributing components: SS_A, SS_B, and SS_{AB}, along with the corresponding partitioning of the degrees of freedom. Mean squares are obtained by dividing the sums of squares by their degrees of freedom. Values for these degrees of freedom are $df_A = a - 1$; $df_B = b - 1$; $df_{AB} = (a - 1)(b - 1)$. Note that these degrees of freedom sum to $df_{between} = ab - 1$.

F-ratios for each of MS_A, MS_B, and MS_{AB} are obtained by dividing the mean square by MS_{within}. The three null hypotheses are tested by comparing these obtained F-ratios with the critical F-values for a given significance level obtained from the table of the F-distribution. Note that each of the three mean squares will have its own F-ratio. Decisions about the three null hypotheses determine which parameters of the full model—all of them, none, or any combination of them—should be retained.

What You Should Know and Be Able to Do

■ Identify (from a verbal description) an independent groups design with any number of conditions (or groups).

■ Given the value of SS_e (SS_{within}), calculate MS_e (MS_{within}).

■ Use MS_e to calculate simultaneous confidence intervals (Tukey's method) for any of the comparisons and explain what the term "simultaneous" means.

■ Calculate Tukey's HSD and explain the difference between planned and post hoc tests of comparisons.

■ Use either simultaneous confidence intervals, or the value of Tukey's HSD, to decide which comparisons are significantly different from zero.

■ Use, when appropriate, Fisher's LSD to decide which comparisons are significantly different from zero.

- Given the description of a between-groups design (either one-way or factorial) and given the number of participants in each condition or group (n), set up an analysis of variance summary table with headings and the degrees of freedom.

- Given the values of the sums of squares, complete the entries in the analysis of variance summary table by calculating the mean squares.

- Use the mean squares to form appropriate F-ratios and compare the values of these ratios with the corresponding critical values found in the table of the F-distribution.

- Use the results of these comparisons to decide which terms of the full model should be retained.

- Use the values in the analysis of variance summary table to calculate R^2.

Short-Answer Questions

1. Will simultaneous confidence intervals be wider or narrower than individual confidence intervals calculated on the same data? Give reasons for your answer.

2. An experiment with an independent-groups design has four conditions. How many different comparisons are there? Why will the confidence intervals for all these comparisons have the same width? On what assumption is this equality based?

3. If everything else is held constant, why would you expect the width of simultaneous confidence intervals to increase as the number of conditions increases?

4. On what grounds are simultaneous confidence intervals usually considered preferable to individual confidence intervals?

5. What is the relationship between the value of Tukey's HSD and the width of simultaneous confidence intervals?

6. If testing comparisons in an experiment with an independent-groups design with more than two conditions, when is it inappropriate to use Fisher's LSD?

7. Explain the relationships among SS_{model}, SS_{total}, $SS_{between}$, and SS_{within}.

8. Explain the distinction between a main effect and an interaction. What does it mean to say that an interaction reflects the non-additivity of the main effects?

9. Explain the relationship between $SS_{between}$ and the sum of squares for the two main effects and the interaction in a two-way factorial design.

10. If the cell means from a factorial design are plotted, what feature of the plot indicates additivity?

Short Problems

Problem 1 An experiment has five conditions and uses a simple independent-groups design with 7 participants in each group.

a. What is the tabled value of q you would use to calculate Tukey's HSD criterion with $\alpha = .05$?

b. If the experimenter had chosen to calculate Fisher's LSD instead of Tukey's HSD, what statistic would replace q, and what is its value?

    c.   Explain why Fisher's LSD is smaller than Tukey's HSD.

Problem 2 An experiment has four conditions (A_1, A_2, A_3, and A_4) and uses a simple independent groups design with 16 participants in each condition. The sum of squares within conditions (SS_e) was 210.0. The means for the four conditions are as follows: A_1: 15.2, A_2: 10.9, A_3: 7.1, A_4: 11.9.

    a.   What is the tabled value of q you would use to calculate Tukey's HSD criterion with $\alpha = .05$?

    b.   What is the value of Tukey's HSD ($\alpha = .05$)? Use this value to calculate a 95% confidence interval for the difference between conditions A_1 and A_2 and between conditions A_1 and A_4.

    c.   The four conditions generate six possible comparisons. Using the obtained value of HSD as your criterion, which of these comparisons is significantly different from 0?

    d.   What is the width of the 95% simultaneous confidence intervals for these six comparisons?

Problem 3 An experiment has three conditions (A_1, A_2, and A_3) and uses a simple independent-groups design with 12 participants in each condition. The means for the three conditions are as follows. A_1: 7.0, A_2: 12.0, A_3: 8.0. Fit the model $Y_i = \mu + \alpha_i + e$ to these results, then calculate $SS_{between}$. The sum of squares within conditions (SS_e) was 495.0. Use this value to complete an analysis of variance summary table for this experiment. Do the results justify the use of Fisher's LSD to test comparisons between pairs of conditions? If they do, then calculate LSD.

Problem 4 An experiment, conducted to evaluate the effect of study method on test performance, has five conditions (five different study methods). A total of 75 volunteers participate in the experiment. Each participant is assigned randomly to one of the five conditions such that there is an equal number (15) in each condition. The total sum of squares is found to be 195.0 and the sum of squares within conditions was found to be 162.0.

    a.   What is the mean square for between conditions?

    b.   What is the mean square for within conditions?

    c.   What is the critical value of the F-ratio (for $\alpha = .05$) you would use to decide which model should be accepted for these results: The null hypothesis model $Y_i = \mu + e$, or the full model $Y_i = \mu + \alpha_i + e$?

Problem 5 An experiment, conducted to evaluate a new anti-depressant drug, uses a design consisting of a control (placebo) condition and three additional conditions: the new drug and two older drugs already in use. Each participating subject is assigned randomly to one of the four conditions such that there is an equal number of subjects in each group. Work out the missing entries in the following analysis of variance summary table.

Source	SS	df	MS
Between conditions	___	___	16
Within conditions	___	___	4
Total	192	___	

What is the *critical* (tabled) value of the F-ratio (for $\alpha = .05$) you would use to decide which model should be accepted for these results: The null model $Y_i = \alpha + e$, or the full model $Y_i = \mu + \alpha_i + e$? By calculating the appropriate F-ratio, decide which of these two models should be accepted in this experiment.

Problem 6 An experiment is conducted with 18 subjects in each of 3 conditions. The response measure is body weight measured in pounds. Analysis of variance of the one-way completely randomized design yields a between-conditions mean square of 15.4 and a within-conditions mean square of 5.6. The experimenter is told to convert the measures from pounds to metric (kilograms) by dividing each score by 2.2. When thus transformed, what will be the value of the

a. mean square for between-conditions.

b. mean square for within-conditions.

c. F-ratio?

Problem 7 A certain drug is known to influence learning in rats, but an experimental psychologist claims that the drug's degree of influence is highly dependent on the animal's level of motivation. The psychologist's claim is that, although it is well known that learning is better with higher levels of motivation, this effect of motivation is much greater under the drug. To test this claim, animals are required to learn to navigate a maze. Two levels of motivation are used (high and low) and each of these conditions is combined with either the drug or a placebo injection. There are thus 4 conditions in all. Nine animals are assigned randomly and independently to each of these conditions.

a. Write out a full model for this experiment.

b. What component of the model is most relevant to the experimenter's prediction?

c. What is the critical (tabled) value of the test statistic you would use to test the null hypothesis most relevant to the experimenter's prediction at the .05 level?

d. Given the sums of squares in the table on page 63, complete the analysis of variance summary table.

e. Which components of your model do the data justify retaining?

	SS	df	MS
Drug (D)	11	——	——
Motivation (M)	25	——	——
D × M	33	——	——
Within conditions	145	——	——

f. Do the data confirm the experimenter's prediction?

Problem 8 In an experimental study of the effect of frustration on aggressive behavior, the conditions consist of three levels of frustration, each paired with two levels of delay between the frustrating event and the measurement of aggression; aggression is measured either immediately or after a 1-hour delay. A total of 48 volunteers are used; they are assigned randomly and independently among the six conditions, 8 per condition. For $\alpha = .05$, what are the critical values of the F-ratio for testing the

a. main effect of frustration.

b. main effect of delay.

c. interaction between frustration and delay?

The analysis of the data yields the sums of squares (SS) shown in the following table.

Source	SS	df	MS	F
Model		——		
Frustration (F)	16.8	——	——	——
Delay (D)	9.8	——	——	——
F × D	6.1	——	——	——
Within conditions (residual)	83.1	——	——	
Total				

Fill in the values for the columns df, MS, and F. Using a Type-1 error rate of $\alpha = .05$, decide on the model that best fits the data. Is this model additive?

Problem 9 In an experimental study of the relationship between fatigue and task difficulty, the conditions consist of three levels of fatigue (low, medium, and high), each paired with two levels of task difficulty, giving six conditions in all. A total of 72 participants are used; they are assigned randomly and independently among the conditions, such that there is an equal number of subjects in each condition. A research assistant analyzes the data using analysis of variance but treats the experiment as if it were a one-way design with 6 conditions. This one-way analysis gives a value of $SS_{total} = 180.1$ and a between-conditions sum of squares of $SS_{between} = 42.1$.

The experimenter points out to the assistant that this is not the intended analysis and that it should have been analyzed as a two-way factorial design. The research assistant performs this intended analysis, obtaining a value of $MS_A = 6.05$ as the mean square for the fatigue factor, and a value of $MS_B = 10.8$ as the mean square for the difficulty factor. Using whatever of the above information is relevant, work out the obtained and the critical F-ratios (for $\alpha = .05$) for main effects and the interaction in the factorial design analysis.

Problem 10 The four tables below contain hypothetical population means from four different experiments, each of which used a two-way factorial design. The two factors have been labeled A (rows) and B (columns). Thus Experiment 1 has 2 levels of A and 3 levels of B.

For each of these four experiments decide which of the effects A, B, or A×B exist.

		B		
Experiment 1	**A**	6.1	4.7	7.5
		6.1	6.1	6.1

		B	
		4.4	4.4
Experiment 2	**A**	5.9	5.9
		4.4	4.4

		B		
		3	5	9
Experiment 3	**A**	7	9	1
		4	6	7

		B			
Experiment 4	**A**	4	6	5	9
		1	3	2	6

Exercises in the Analysis of Realistic Data

Set 1: Understanding Memory

Exercise 1. Further test of the generation effect (Data Set MEM_06.dat)

In Exercise 1 of Chapter 6 you analyzed the data set MEM_02.dat and found that the "generation" condition produced better remembering than the "read" condition. A critic of that experiment points out that this difference may have nothing to do with the different kind of mental processing that the two tasks require but instead may be a consequence of a cue word being present in the case of the generate condition but not in the read condition. Perhaps the superiority of the generate condition stems from the association formed between the target word and the cue word.

To test this possibility the experiment is expanded to include a third condition called a "read-pair" condition. Participants in this condition see both the cue word and the target word (large - small, for example) and are instructed to look at both words but say out loud only the right-hand word. In this way all three groups spoke an identical set of words, participants in the read and read-pair conditions by reading them directly, participants in the generate group by generating the target word in response to the cue word. Following the presentation of the words, all participants were given a recognition test of the 40 spoken words. There were 54 participants; 18 were randomly assigned to each condition. The response variable is the percentage of words correctly recognized.

Data Set MEM_06.dat.

Generate: $\bar{Y}_1 = 80.22$; $SS_1 = 1275.1$

79	69	76	86	79	97	92	73	87
74	81	63	75	81	92	76	87	77

Read-Pair: $\bar{Y}_2 = 68.61$; $SS_2 = 1532.3$

60	71	61	62	71	52	72	62	63
73	51	75	81	79	78	77	82	65

Read: $\bar{Y}_3 = 53.56$; $SS_3 = 1276.4$

36	40	50	53	64	43	62	45	52
57	65	53	65	50	55	52	60	62

$SS_{total} = 10,519.4$

1. Plot a bar chart of the mean recognition scores for the three conditions, adding error bars indicating the standard error of these means.

2. Calculate Tukey's HSD and use it to obtain 95% confidence intervals for the three comparisons between the means.

3. Do the results suggest the need for any change in the interpretation of the experiment analyzed in Exercise 1 of Chapter 6 (Data Set MEM_02.dat)?

4. Calculate the value of R^2 for this experiment.

5. A reviewer of the report of this experiment requests that you report the results of an analysis of variance and an overall *F*-test of the null hypothesis. Although you argue (correctly) that such an analysis adds nothing important to the understanding of the results, you agree to do so. Write out the analysis of variance summary table and conduct the relevant *F*-test. State explicitly the null hypothesis that this *F*-test evaluates.

Exercise 2. A test of the generation effect with a different response measure (Data Set MEM_07.dat)

The experiment in Exercise 1 is repeated but in this version the recognition test is replaced with a measure of "implicit memory." This test consists of presenting words in a brief flash such that they cannot always be identified. However, it is well known that a recent prior exposure to a word facilitates its identification under these conditions, a phenomenon known as priming. Priming is assumed to reflect the effect of implicit memory. Implicit memory involves the influence of a past experience without the conscious intent to use this experience, or even having any awareness of its influence. A recall or recognition test, on the other hand, is assumed to involve explicit memory in that such tests are preceded by direct instructions to "try to remember." The experiment in this exercise addresses the question of whether implicit memory will display the generation effect in the same way as is observed with recognition memory.

Following the presentation of the words, all participants were given the identification test of the spoken words. There were 54 participants; 18 were randomly assigned to each condition. The response variable is the percentage of words correctly identified.

Data Set MEM_07.dat

Generate: $\bar{Y}_1 = 53.72$; $SS_1 = 1027.6$

38	51	58	47	59	49	59	51	53
54	65	54	61	53	49	39	61	66

Read-Pair: $\bar{Y}_2 = 68.22$; $SS_2 = 1493.1$

63	73	61	57	78	50	71	68	61
71	54	70	80	84	71	75	78	63

Read: $\bar{Y}_3 = 77.61$; $SS_3 = 1320.3$

74	68	74	84	69	88	90	81	86
72	79	57	82	76	89	67	79	82

$SS_{total} = 9055.5$

1. Plot a bar chart of the mean recall scores for the three conditions, adding error bars indicating the standard error of these means.

2. Calculate Tukey's HSD and use it to obtain 95% confidence intervals for the three comparisons between the means.

3. Calculate the value of R^2 for this experiment.

4. Do the results indicate differences between implicit and explicit memory with respect to the generation effect?

Exercise 3. Levels of processing (Data Set MEM_08.dat)

In this version of an experiment demonstrating the principle of "levels of processing" in memory, participants were presented with 40 different to-be-remembered words. Each word was presented just once, preceded by a question. The question could be one of three types, these question types constituting the three conditions of the experiment. There were 75 participants who were assigned randomly to the conditions with $n = 25$ participants in each. Using the word "flame" as an example, the question types (conditions) were:

1. *Appearance:* Is the word typed in capital letters?   **FLAME**

2. *Sound:* Does the word rhyme with "claim"? **FLAME**

3. *Meaning:* Is the word something hot? **FLAME**

For the given examples, the correct answer to each of these questions is "yes." For other questions the correct answer was "no." After this encoding phase, participants were given an unexpected recall test. The response variable was the number of words correctly recalled. The following data are for words to which, in the encoding phase, the correct answer was "yes."

Appearance: $\bar{Y}_1 = 6.32$, $SS_1 = 381.44$

4	2	2	14	13	8	10	6	6	5	14	4	1
4	2	6	3	4	7	9	12	4	9	1	8	

Sound: $\bar{Y}_2 = 10.08$; $SS_2 = 409.84$

12	14	9	7	7	10	10	16	5	2	4	10	6
11	9	19	4	10	8	14	12	11	15	13	14	

Meaning: $\bar{Y}_3 = 14.28$; $SS_3 = 309.04$

12	15	20	18	13	18	8	12	20	12	17	17	16
16	13	20	16	11	11	14	14	12	14	6	12	

$SS_{total} = 1893.15$

1. Plot a bar chart of the mean recall scores for the three conditions, adding error bars indicating the standard error of these means.

2. Write out the analysis of variance summary table and conduct the relevant F-test. If justified, calculate Fisher's LSD for the three comparisons in this experiment. What conclusions would you draw about differences among the three conditions?

3. Calculate the value of R^2 for this experiment.

Exercise 4. Levels of processing in implicit memory (Data Set MEM_09.dat)

Exercises 1 and 2 investigated the differences between implicit and explicit memory with respect to the generation effect. A similar question might be asked with respect to levels of processing. Does implicit memory show a levels-of-processing effect? To answer this question, the experiment described in Exercise 3 is repeated with a change in the response measure and 21 participants in each condition. The test is now the implicit memory test used in the Exercise 2 experiment—the identification of briefly flashed words. The response measure is the proportion of words (out of 40) correctly identified. $SS_{total} = 0.743$

Appearance: $\bar{Y}_1 = .645$, $SS_1 = 0.156$

.650	.600	.750	.700	.775	.700	.650	.675	.550	.650	.700
.600	.625	.525	.500	.675	.625	.725	.525	.525	.825	

Sound: $\bar{Y}_2 = .674$; $SS_2 = 0.309$

.875 .450 .625 .675 .600 .625 .750 .625 .825 .575 .625

.575 .725 .800 .875 .775 .650 .650 .675 .400 .775

Meaning: $\bar{Y}_3 = .707$; $SS_3 = 0.238$

.775 .700 .750 .475 .775 .600 .750 .550 .625 .925 .625

.600 .750 .700 .800 .875 .600 .775 .675 .750 .775

1. Plot a bar chart of the mean recall scores for the three conditions, adding error bars indicating the standard error of these means.

2. Write out the analysis of variance summary table and conduct the relevant F-test. If justified, calculate Fisher's LSD for the three comparisons in this experiment. What conclusions would you draw about differences among the three conditions?

3. Suppose the response measure was converted from proportion to number recognized by multiplying each proportion by 40. Write out the revised analysis of variance summary table. Why is the value of the F-ratio unchanged?

4 Calculate the value of R^2 for this experiment.

5. Does implicit memory show a levels-of-processing effect?

Exercise 5. Levels of processing with incidental instructions (Data Set MEM_10.dat)

This experiment investigated whether the levels-of-processing effect is influenced by participants being given prior instructions that they will be asked to recall the words they are about to see. The three conditions used in Exercises 3 and 4 were used, but now a randomly chosen half of the participants were forewarned (intentional instruction condition) about the recall test. The other half (incidental instruction condition) were given no such warning. The experiment therefore has six conditions arranged in a 3 (levels) by 2 (incidental/intentional) design. A total of 90 participants were used, 15 being randomly assigned to each condition.

1. Write out the full and the null model for this experiment.

2. Complete the entries in the following analysis of variance summary table and conduct the relevant F-tests.

	Source	SS	df	MS	F-ratio
Model:	Between conditions				
	A (Level)	1876.1			
	B (Instruction)	17.8			
	A × B	11.1			
Residuals:	Within conditions				
Total		3150.2			

3. Calculate the value of R^2 for this experiment.

4. How would you describe the effect of incidental versus intentional instructions on the levels-of-processing effect?

Set 2: Child Behavior

Exercise 6. Perspective taking in parents (Data Set CHILD_2.dat)

Data from this experiment were examined Exercise 6 of Chapter 2, and Exercise 4 of Chapter 3. We can now complete the analysis. What factors influence whether or not parents punish their children for misbehavior? An obvious factor is the nature of the misbehavior. Another possible influence is the extent to which parents view the misbehavior from their own perspective or from that of the child. This study investigated this possibility using three conditions. A total of 45 mothers participated. They were assigned randomly to the three conditions, 15 in each. All mothers read a story about a child's misbehavior.

The mothers in **Condition 1** were instructed to "think about the misbehavior from the child's point of view. What was the child thinking and feeling at the time when the behavior occurred?"

The mothers in **Condition 2** were instructed to "think about the misbehavior from the your point of view. What would you be thinking and feeling at the time when the behavior occurred?"

The mothers in **Condition 3** formed a control condition and were given no special instructions.

After reading the story all parents used a seven-point scale to rate the likelihood that they would punish the child for the misdeed. In chapter 2, you found the means to be 2.6, 3.9, and 4.0 for conditions 1, 2, and 3 respectively. In Chapter 3 you found $SS_{between} = 18.71$ and $SS_{total} = 85.24$.

1. Plot a bar chart of the mean recall scores for the three conditions, adding error bars indicating the standard error of these means.

2. Calculate Tukey's HSD and use it to obtain 95% confidence intervals for the three comparisons between the means.

3. Calculate the value of R^2 for this experiment.

4. Do the results indicate differences between the conditions?

Exercise 7. Possible differences in scholastic aptitude (Data Set TEST_3.dat)

In this exercise you return to your role as the Director of Psychological Services for a large school board who has constructed a scholastic aptitude test. You are asked to answer the question as to whether there are any differences in scholastic aptitude among the four major high schools within your district. To answer this question the scholastic aptitude test is administered to a random sample of 100 students from each of the four schools.

Stemplots of the scores for four schools are shown below. Use the given values of Q_1, the median, and Q_3 to draw side-by-side boxplots. Do these plots suggest any differences among the schools?

Group 1: $\bar{Y}_1 = 85.86$;

$Q_1 = 64.5$; Md = 86.0; $Q_3 = 105.75$

3	122446
4	4679
5	001445599
6	3333446999
7	123333899
8	11234455566677777
9	00123358
10	000111223456779
11	00113333789
12	11346
13	1
14	115
15	2

Group: 2: $\bar{Y}_2 = 85.96$

$Q_1 = 65.25$; Md = 86.5; $Q_3 = 106.0$

2	8
3	168
4	0024688
5	01146678
6	1113356899
7	0012445889
8	0012344466677789
9	00134455667
10	00122334668
11	0023466789
12	112455568
13	13
14	45

Group 3: $\bar{Y}_3 = 84.17$

$Q_1 = 62.25$; Md = 80.0; $Q_3 = 109.25$

1	3
2	
3	03678
4	1356678
5	22222558
6	011233447788
7	012345577788899
8	0001445566889
9	166778999
10	00457
11	004466668889
12	12344689
13	15
14	7
15	15

Group 4: $\bar{Y}_4 = 88.01$

$Q_1 = 70.0$; Md = 86.0; $Q_3 = 109.0$

1	9
2	06
3	
4	1459
5	2223777789999
6	1179
7	001133446788999
8	00111455666688
9	0122234455679
10	0012344699
11	011223677788
12	12
13	0122355
14	1
15	
16	25

1. Given the result that $MS_e = 804.0$, plot a bar chart of the mean scholastic aptitude test scores for the four schools, adding error bars indicating the standard error of these means.

2. Calculate Tukey's HSD for comparisons between pairs of schools and decide whether the data support any claim to there being differences among the schools.

Exercise 8. Possible sex differences in scholastic aptitude (Data Set TEST_4.dat)

A member of your school board asks whether there is a sex difference in scholastic aptitude in any of the four schools studied in Exercise 6. To answer this question you obtain scholastic aptitude test scores for a random sample of 50 boys and 50 girls from each of the four schools. The resulting means are given in the following table.

	School			
	B_1	B_2	B_3	B_4
(A1) Female	$\bar{Y}_{11} = 86.16$	$\bar{Y}_{12} = 90.32$	$\bar{Y}_{13} = 86.02$	$\bar{Y}_{14} = 92.98$
(A2) Male	$\bar{Y}_{21} = 85.18$	$\bar{Y}_{22} = 83.74$	$\bar{Y}_{23} = 84.48$	$\bar{Y}_{24} = 83.88$

1. Draw a graph of these results. Does there appear to be an interaction between gender and school?

2. Write out the full and the null model for this experiment.

3. Complete the following analysis of variance summary table and conduct the relevant F-tests.

Source		SS	df	MS	F-ratio
Model:	Between conditions:		—		
	A (Gender)	2,070.3	—	—	—
	B (School)	622.1	—	—	—
	A × B	1,165.7	—	—	—
Residuals:	Within conditions:	27,4354.3	—	—	
Total					

4. Calculate the value of R^2 for this experiment.

5. Is there any evidence for sex differences in scholastic aptitude among any of the four schools?

8 INCREASING THE PRECISION OF AN EXPERIMENT

Overview

Chapter 8 addresses the question of how to evaluate and improve the precision of an experiment. The precision of an experiment can be increased either by increasing the size of n, or by lowering the value of MS_e.

1. Estimating sample size. The precision of an experiment can be measured in terms of w, the half-width of the confidence interval. It is possible to estimate a value of n that will yield an approximation to some desired value of w. Precision can also be measured in terms of power $(1 - \beta)$ where β is the Type-2 error rate. The sample size chosen to achieve a given value of w will give an approximate power level $(1 - \beta)$ of .5 to detect a true difference of w. Using appropriate tables, it is possible to estimate a value of n needed to yield an approximation to any desired level of power $(1 - \beta)$ relative to a designated effect size, **d**.

2. Reducing MS_e by matching. A strategy for reducing MS_e and thereby increasing precision is to form matched pairs. Matching will be effective in reducing MS_e to the extent that pairs are matched on a factor that is responsible for variability among participants *within* a condition. A simple form of analysis for matched pairs is to take difference scores, calculate the standard error of their mean, and use the t-distribution to calculate a confidence interval or to evaluate the statistical significance of $\overline{D}$ directly.

The full model for the matched-pair design is $Y = \mu + \alpha_i + \pi_j + e$. This model forms the basis of an analysis of variance in which SS_{within} of the completely randomized design is partitioned into two components: a reduced SS_{within} and SS_{pairs}. It is this removal of the SS_{pairs} that is responsible for the increased precision. An experimental design that can be thought of as a special case of matching is known as within-subjects. Although the analysis of within-subjects designs is usually performed in exactly the same way as matched pairs, care must be taken with within-subjects designs to ensure that results are not contaminated by unwanted carryover effects.

3. Randomized block and larger within-subjects designs. The matched-pair design can be extended to experiments with more than two conditions. If the experiment uses blocks of matched participants, then the design is described as a randomized block design. An experiment with k conditions would consist of n blocks, each made up of k matched participants. If the experiment obtains an observation from each participant under all of the k conditions then the experiment is described as a within-subjects or repeated measures design.

As in the two-condition case, both randomized block and within-subjects (repeated measure) designs are analyzed in the same way. Moreover, both the model and the analysis of variance summary table have exactly the same form as in the two-condition case. When there are three or more conditions the problem of multiple confidence intervals arises as it did in completely randomized designs. The solution to this problem is the same as for completely randomized designs. The Studentized range statistic can be used to obtain simultaneous confidence intervals and to obtain Tukey's HSD.

What You Should Know and Be Able to Do

■ Describe the two general strategies for increasing the precision of an experiment: by increasing the size of n, or by lowering the value of MS_e.

■ Estimate a value of n that will yield an approximation to some desired value of w.

■ Use published tables to estimate a value of n needed to yield an approximation to any desired level of power $(1 - \beta)$ relative to a designated effect size, **d**.

■ Explain how the appropriate use of matched pairs reduces the size of MS_e.

■ Calculate a confidence interval for the difference between the two means of a matched pair design.

■ Explain the meaning of the terms in the full model for the matched-pair design:

$$Y = \mu + \alpha_i + \pi_j + e$$

■ Explain the close correspondence between matched-pairs or randomized block designs and within-subjects designs.

■ Explain how carryover effects are the most important potential difference between matched-pairs or randomized block designs and within-subjects designs.

■ Set out the components of an analysis of variance summary table for a within-subjects, matched pair, or randomized block designs, calculate degrees of freedom, mean squares, and F-ratios.

■ Calculate simultaneous confidence intervals for the difference between pairs of means in randomized block and within-subjects designs.

Short-Answer Questions

1. State two ways of describing the degree of precision of an experiment.

2. State two ways of increasing the precision of an experiment.

3. If an experimenter increases the sample size (n) in an experiment, which (if any) of the following values would be increased: w, β, 1-β?

4. In a study of the relative effectiveness of two methods of teaching algebra, the

investigator forms matched pairs, assigning one member of each pair to one method, the other to the second method. Two bases for matching are considered: (a) Score on the Math section of the SAT (b) Score on the Verbal section of the SAT. Which of these two is likely to result in the greater increase in precision? State why.

5. Explain what is meant by a carryover effect. Think of an example of a repeated-measures design that you think would probably produce a carryover effect.

6. Explain the sense in which a repeated-measures design can be thought of as a special case of a randomized block design.

7. For a within-subjects design, which combination of the following sums of squares adds up to form SS_{model}: SS_e, $SS_{between}$, or $SS_{subjects}$?

8. Suppose that in a randomized block design, MS_{blocks} is approximately the same value as MS_e. What does this result suggest about the variable on which the blocking was based?

Short Problems

Problem 1 An investigator wishes to estimate the difference in simple reaction time between elderly (aged 75-80) and middle aged (aged 45-50) populations to within 50 milliseconds in the sense of obtaining a 95% confidence interval with a total width ($2 \times w$) of 100 milliseconds. Previous research has indicated that the standard deviation of reaction times for both these age groups is 150 milliseconds. Assuming an equal number in each age group, how many participants does the investigator need in each group to achieve this goal?

Problem 2 An experiment is designed to test whether degree of illumination (dim versus bright) influences judged length of a cylindrical rod. Given an estimated standard deviation of 1.5 centimeters for such judgments under both of these conditions, how many participants are needed in each of two independent groups to achieve a power of .7 to detect a true difference of one centimeter between the two conditions?

Problem 3 An investigator decides to conduct the experiment described in Problem 2 as a within-subjects design. How might this experiment avoid carryover effects biasing the results? If the pilot work has established an estimate of 1.1 centimeters as the standard deviation of the difference between judgments in the two conditions, how many participants are needed to achieve a power of .7 to detect a true difference of one centimeter between the two conditions?

Problem 4 A school board decides that it is worth adopting a novel method of instruction if it improves performance on a final exam by at least 5%. In an experiment designed to compare the novel method with the traditional one, students are to be assigned randomly to one of the two methods. Given an estimate of 9.0% as the standard deviation of final exam scores, how many students are needed in each group to ensure a 95% confidence interval with a width ($2 \times w$) of 10%? What is the power of this experiment to detect a true difference of (a) 5%? (b) 3%?

Problem 5 An experiment uses 32 sibling pairs, assigning one sibling from each pair to each of the two conditions.

a. What is the tabled value of the t-statistic you would use to calculate the 95% confidence interval for the difference between the true means of these conditions?

b. The estimated standard error of the difference between the means of these two conditions was found to be 2.3. Calculate the width of the confidence interval for the true difference between the means of these two conditions.

c. What is minimum difference between the observed means of these two conditions needed to reject the null hypothesis with $\alpha = .05$, two tailed?

d. Using 2.3 as the value of the standard error of the difference between the means, what sample size would be needed to achieve a power of .7 to detect a true mean difference of 4.0?

Problem 6 Three different two-condition experiments were conducted using a within-subjects design. In the following summary of each experiment, the value of n is the number of pairs of observations in that experiment and the variance is the variance of the n difference scores for that experiment. Calculate a 95% confidence interval for the difference between the means for each experiment.

a. Experiment A ($n = 33$) has a mean difference of 4.8 and a variance of 169.0.

b. Experiment B ($n = 7$) has a mean difference of 3.2 and a variance of 23.

c. Experiment C ($n = 40$) has a mean difference of 4.1 and a variance of 91.

Problem 7 Three different experiments were conducted using a matched-pairs design. In the following summary of each experiment, the value of n is the number of pairs of observations in that experiment. The stated variance is the variance of the n difference scores for that experiment.

If the Type-1 error rate is set at $\alpha = .05$, decide which model should be accepted for each experiment: the null hypothesis model $Y_i = \mu + \pi_j + e$, or the full model $Y_{ij} = \mu + \alpha_i + \pi_j + e$.

a. Experiment A ($n = 25$) has a mean difference of 2.8 and a variance of 25.0.

b. Experiment B ($n = 9$) has a mean difference of 2.9 and a variance of 81.0.

c. Experiment C ($n = 25$) has a mean difference of 1.5 and a variance of 1000.

Problem 8 As part of a research program in reading, an experiment investigates whether visual word recognition of briefly flashed words is influenced by word length, and if so, by how much. Fifteen participants each see a randomly ordered sequence of 60 common words, 20 of which are short, 20 medium, and 20 long. The response measure is the proportion of words correctly identified for each word length.

The means for the three conditions were .6, .5, and .4. The total sum of squares was 6.274 and the sum of squares for participants was 4.96. Calculate $SS_{between}$ and then write out the complete analysis of variance summary table for this experiment, evaluate the null hypothesis that word length has no effect, and obtain simultaneous 95% confidence intervals for the differences among pairs of conditions.

Exercises in the Analysis of Realistic Data

Set 1: Understanding Memory

Exercise 1. Type-2 error in memory for chess positions (Data Set MEM_04.dat)

Exercise 3 of Chapter 6 compared the memory of expert and novice chess players for chess pieces arranged in random positions on the board. It found no evidence for such a difference. A critic points out that because the null hypothesis was retained, a Type-2 error might have been made. Address this criticism by evaluating (in whichever way you consider appropriate) the power and precision of that experiment. The results of basic computations were as follows.

	n	Mean	SS	SS_{total}
Experts	28	7.75	131.25	300.55
Novices	28	6.93	159.86	

Exercise 2. Test the generation effect using a different design (Data Set MEM_11.dat)

In Exercise 1 of Chapter 6 you analyzed the data set MEM_02.dat and found that the "generation" condition produced better remembering than the "read" condition. The version for this exercise uses a within-subjects design. All participants ($n = 25$) saw a list of common words such as "small" and "happy." They were instructed that if the word had an asterisk to its left, then they were to read it out loud in preparation for a subsequent recall test. If they saw a word along with the first letter of a second word, they were instructed to generate a word that started with this letter and meant the opposite of that word, and that they would be asked to recall the generated word. The order in which the words were presented was random. Following the presentation of the words, all participants were asked to recall as many of the spoken words as they could remember. The number of words recalled in each condition by the 25 participants is contained in the data set MEM_11.dat.

The means were 10.52 and 13.20 words recalled for the read and generate conditions respectively. The analysis of variance summary table for this experiment is shown below. Complete the missing entries.

Source	SS	df	MS
Model	——	——	
Between conditions	89.78	——	——
Participants	270.52	——	——
Residuals	31.72	——	——
Total			

a. Calculate the value of the standard error of the difference between the means.

b. Describe the effect size in this experiment by calculating an estimate of Cohen's **d**.

   c. Calculate a 95% confidence interval for the difference between the means of the two conditions.

   d. Calculate R^2, the percentage reduction in the sum of squared residuals achieved by the full model compared to the null model.

   e. Do the data justify the full model or is the simpler null model an adequate account?

   f. What conclusion would you draw about the effect of the experimental manipulation?

   g. Compare the precision of this experiment with that obtained in the previous version of the experiment using an independent-groups design (Exercise 1 of Chapter 6; Data Set MEM_02.dat).

Exercise 3. An extended test of the generation effect using a different design (Data Set MEM_12.dat)

In Exercise 1 of Chapter 7 you analyzed Data Set MEM_06.dat from an extended three-condition version of the generation effect experiment. See the description there for details of the three conditions. This new version of the three-condition experiment uses a within-subjects design. All participants ($n = 18$) saw a list of common words, with an equal number of words in each condition. The order in which the words were presented was random. After this encoding phase, participants were given a recognition-memory test. For this test phase, the words studied in the encoding phase were mixed with a number of previously unseen words to serve as distracters, and participants were required to identify the words that had been previously presented. The percentage of words recognized by each participant is contained in Data Set MEM_12.dat. The mean recognition scores were Read = 54.8, Read-pair = 69.1, Generate = 80.1.

a. Write out the full model for this experiment.

b. Complete the following analysis of variance summary table of the results and conduct the relevant F-test for evaluating the null hypothesis of no differences among the conditions.

Source	SS	df	MS
Model	——	——	
Between conditions	5,809.3	——	——
Participants	4,224.7	——	——
Residuals	1,416.0	——	——
Total			

c. Calculate Tukey's HSD and use it to obtain 95% confidence intervals for the three comparisons between the means.

d. Do the results constitute a general replication of those obtained in the previous version of this experiment using an independent groups design (Exercise 1 of Chapter 7; data set MEM_06.dat)?

e. Calculate the value of R^2 for this experiment.

f. Compare the precision of this experiment with that obtained in the previous version of the experiment using an independent-groups design (Exercise 1 of Chapter 7; data set MEM_06.dat).

g. Do you think this version of the experiment might suffer from carryover effects?

Exercise 4. Investigation of levels of processing using a different design (Data Set MEM_13.dat)

In Exercise 3 of Chapter 7 you analyzed data set MEM_08.dat from an experiment investigating the levels of processing effect. See the description there for details of the three conditions. This new version of the experiment uses a within-subjects design. All ($n = 16$) participants saw a list of common words, each preceded by one of the three questions, with an equal number of words for each question. The order in which the words were presented was random.

After this encoding phase, participants were given a recognition-memory test. For this test phase, the words studied in the encoding phase were mixed with a number of previously unseen words to serve as distracters, and participants were required to identify the words that had been previously presented. The three response measures for each participant were the percentage of words correctly identified as having been previously presented for each of the three question types. The data are for words to which, in the encoding phase, the correct answer was "yes."

Data Set MEM_13.dat

P	App.	Sound	Meaning
1	35	51	70
2	34	59	66
3	49	57	74
4	65	73	89
5	51	62	66
6	44	48	81
7	55	68	83
8	47	59	80
9	60	70	88
10	63	80	79
11	68	57	83
12	42	63	58
13	62	82	83
14	53	80	78
15	52	75	83
16	29	53	56
Means	50.56	64.81	76.06

a. Complete the following analysis of variance summary table of the results

b. Conduct the relevant *F*-test for evaluating the null hypothesis of no differences among the conditions.

Source		SS	*df*	MS
Model		——	——	
	Between conditions	5,226.0	——	——
	Participants	4,003.3	——	——
Residuals		——	——	——
Total		10,559.3		

c. Calculate Tukey's HSD and use it to obtain 95% confidence intervals for the three comparisons between the means.

d. Do the results constitute a general replication of those obtained in the previous version of this experiment using an independent-groups design (Exercise 3 of Chapter 7; Data Set MEM_08.dat)?

e. Calculate the value of R^2 for this experiment.

f. Compare the precision of this experiment with that obtained in the previous version of the experiment using an independent-groups design (Exercise 3 of Chapter 7; Data Set MEM_08.dat).

g. Do you think this version of the experiment might suffer from carryover effects?

Set 2: Child Behavior

Exercise 5. Precision in the mood and compliance experiment (Data Set CHILD_5.dat)

This experiment, analyzed in Chapter 6, asked whether children were more likely to comply promptly to a request if they were in a good mood. A total of 46 seven-year-old children participated in the experiment. They were assigned randomly and independently to either a neutral mood control condition or a gift (good mood) condition, 23 in each. The response measure was the time taken (latency) for the child to begin the task. The sum of squares for the neutral control condition was 359.3 and for the gift condition it was 808.9.

In designing a replication of the experiment the investigator wants to estimate the difference in response time between the two conditions to within 3 seconds in the sense of obtaining a 95% confidence interval with a total width ($2 \times w$) of 6 seconds.

a. Using the results from the earlier experiment, estimate how many participants are needed in each condition to achieve this goal.

b. If the experiment is conducted with this recommended sample size, what is the power of the *t*-test to detect a true difference of 4 seconds?

Exercise 6. Perspective taking in parents (Data Set CHILD_7.dat)

This experiment was described and analyzed in Chapters 2 (Exercise 6), 3 (Exercise 4), and 7 (Exercise 6). This new version of the experiment uses a within-subjects design.

All ($n = 15$) parents read three stories and used a seven-point scale to rate the likelihood that they would punish the child for the misdeed. For each of the three stories they were asked to take a different one of the three perspectives. Thus each parent contributes three ratings, one for each perspective.

Condition 1: "Think about the misbehavior from the child's point of view." Mean = 2.27
Condition 2: "Think about the misbehavior from your point of view." Mean = 3.87
Condition 3: Control condition, no specific instructions. Mean = 3.80

a. Complete the following analysis of variance summary table of the results and conduct the relevant F-test for evaluating the null hypothesis of no differences among the conditions.

Source		SS	df	MS
Model		——	——	
	Between conditions	24.58	——	——
	Participants	46.31	——	——
Residuals		22.76	——	——
Total		——		

b. Calculate Tukey's HSD and use it to obtain 95% confidence intervals for the three comparisons between the means.

c. Do the results constitute a general replication of those obtained in the previous version of this experiment using an independent-groups design (Exercise 6 of Chapter 7)?

d. Calculate the value of R^2 for this experiment.

e. Compare the precision of this experiment with that obtained in the previous version of the experiment using an independent-groups design (Exercise 6 of Chapter 7).

f. Do you think this version of the experiment might suffer from carryover effects?

QUANTITATIVE PREDICTOR VARIABLES: LINEAR REGRESSION AND CORRELATION

9

Overview

Chapter 9 describes the analysis of data for which both predictor and response variables are quantitative. There are two general goals of such a data analysis.

1. Regression. The goal might be to construct a regression equation for predicting values of the response variable. When this is the primary goal the analysis is usually described as a regression analysis. A regression analysis is usually applied to data in which there is a clear distinction between response and predictor variable. In such applications it is quite often the case that values of predictor variables are not a random sample from the population of possible values, but points that have been carefully selected (often at equal intervals) to cover a desired range. Drug dosage levels or number of training sessions constitute prototypical examples.

2. Correlation. The goal might be to describe the strength of the relationship between the two variables. When this is the primary goal the analysis is usually described as a correlational analysis. This analysis is commonly associated with data in which there is no sharp distinction between response and predictor variables. Both can be considered random variables and the roles of response and predictor variables can be interchanged.

3. Scatterplots. Data for which both predictor and response variables are quantitative can be represented graphically in the form of a scatterplot. A scatterplot is a graph that locates each data point in a space defined by its value on the predictor variable (plotted on the x-axis) and its value on the response variable (plotted on the y-axis).

4. Assumptions. The procedures described in this chapter make several assumptions. The first assumption is that the marginal distribution and the conditional distributions of the response variable are samples from normally distributed random variables. A conditional distribution is the distribution of the response variable associated with a specified value of the predictor variable; the marginal distribution is the distribution of all the values of the response variable regardless of value of the predictor variable. For applications in which both variables can be considered random variables and the roles of predictor and response variable interchanged, we assume that the sample data have been drawn from a bivariate normal probability distribution. In such a distribution all the conditional distributions of both variables are normal distributions as are both marginal distributions. The second assumption is that the conditional distributions of the response variable have equal variances (homogeneity of variance assumption). The third assumption is that the means of these conditional distributions fall along a straight line. This is the assumption of linear regression.

5. Fitted regression line. The fitted version of the least-squares linear regression rule is $\hat{Y} = a + bX$. The value a is the intercept of the line and b is the slope. The values of a and b are chosen to minimize the sum of squared residuals. The value of b is known as the estimated regression coefficient or regression weight. The sum of squares of the residuals for the fitted model has $N - 2$ degrees of freedom. The sum of products is defined as $SP_{XY} = \Sigma(X - \bar{X})(Y - \bar{Y})$. This value enters into the formula for the least-squares estimate of the regression weight, b. For the bivariate normal model, two regression lines can be obtained. One is for the regression of Y on X, the other for the regression of X on Y.

6. Standard error of the regression coefficient. The standard error of the estimated regression weight can be calculated and used to obtain a confidence interval for β. This confidence interval can be used in the normal way to evaluate the null hypothesis that the true value of β is zero.

7. Regression analysis. Regression problems can be analyzed by partitioning the total sum of squares according to the model $Y = \alpha + \beta X + e$ where α is the parameter representing the intercept—the point that the regression line crosses the y-axis, and β is the regression coefficient—the slope of the regression line. This regression analysis summary table partitions the total sum of squares into a linear regression sum of squares with 1 degree of freedom, and the residual sum of squares with $N - 2$ degrees of freedom. The symbol N denotes the total number of observations.

The square root of the mean square of the residuals is known as the standard error of estimate. The residual mean square from the regression analysis can be used to form an F-ratio to evaluate the hypothesis that the regression coefficient, β, is zero. The sum of squares for regression can be compared to the total sum of squares to obtain R^2, the proportion of the variability accounted for by the linear regression model. When the data are assumed to have an underlying bivariate normal distribution, $\sqrt{R^2}$ is denoted by r, and is termed the Pearson product-moment correlation coefficient.

8. Examining assumptions. There are several ways in which to examine the validity of the linearity assumption. Graphically, residuals can be plotted against predicted values of the response variable, or against the conditional means of the predictor variable, and examined for systematic trends. More formal methods are available but the only case considered in detail was that in which the predictor variable is manipulated and the regression analysis can be compared to an analysis of variance based on the full model that treats the predictor variable as categorical. When linearity is inappropriate, the linear regression model can give a false picture of the strength of the relationship between the variables.

9. Product moment correlation coefficient. Although the product-moment correlation coefficient is widely used as a descriptive statistic, it is important to interpret it within the wider context of linear regression. The correlation coefficient is a special case of $\sqrt{R^2}$ that is applicable when that data have an underlying bivariate normal distribution. Various formulas for the correlation coefficient demonstrate its relation to the sum of products and the estimated regression coefficient. In particular, r is equal to the regression coefficient multiplied by the ratio of the standard deviations of the marginal distributions.

The size of a correlation coefficient is reflected in the shape of the scatterplot. As the size of the correlation coefficient increases, the overall shape of the scatterplot becomes less circular and more elliptical. Various factors can influence the size of the correlation coefficient. The factors covered in this chapter are selection factors that restrict or inflate variance, outliers, curvilinear regression, and measurement reliability.

What You Should Know and Be Able to Do

■ Distinguish (in terms of the goals of the analysis) between situations typically described as regression from those described as correlational.

■ Describe how to obtain a scatterplot.

■ Explain the difference between conditional and marginal distributions.

■ State the three assumptions underlying regression analyses.

■ Set out the components of a regression summary table, calculate degrees of freedom, mean squares, and F-ratios.

■ Explain what is meant by the terms "intercept" and "slope" when applied to the parameters of a regression line.

■ Explain the similarities and differences between the meaning of the assumption of homogeneity of variance when applied to regression analysis as opposed to an analysis of variance of a independent-groups design.

■ State the effect of violation of the linearity assumption on the value of R^2 (and r, the product-moment correlation coefficient).

■ Describe the relationship between the size of the product-moment correlation coefficient and the shape of the scatterplot.

Short-Answer Questions

1. Measures of IQ and GPA are obtained from a large sample of undergraduate students and the product-moment correlation between the two measures is calculated. An administrator asks the following questions. Decide which questions refer to a marginal distribution, and which to a conditional distribution.

   a. What is the mean GPA of students with an IQ of 120?

   b. What is the mean GPA of all students?

   c. What is the standard deviation of GPA scores of all students?

   d. What is the mean IQ of students with a GPA of 3.0?

2. Assuming there is a significant correlation between IQ and GPA, which of the following distributions would have the smaller variance: The marginal distribution of GPA scores, or the conditional distribution of GPA scores of students with an IQ of 110? Give reasons.

3. A linear regression model is fitted to data and R^2 is found to be .37. However closer inspection of the data shows the regression to be nonlinear. When the more appropriate nonlinear model is fitted, the revised value of R^2 will be (a) unchanged, (b) greater than .37, (c) less than .37, (d) unknown without actually performing the analysis.

4. Decide whether the following statement is true or false. One possible way to increase the correlation between two variables X and Y is to make the sample more homogeneous with respect to X or Y thereby reducing the variance of X or Y (thereby reducing SS_X or SS_Y).

5. An experimenter reports that the sum of products between two measures (Y_1 and Y_2) of spatial reasoning for 100 subjects is 2210, and that the variance of the Y_1 and Y_2 scores is 16 and 25 respectively. How can you tell that the experimenter has made a mistake in these calculations?

6. The product-moment correlation between two measures X and Y is .40. If all the scores for the X measures are divided by five (and the Y measure is left unchanged) what is the effect on the correlation coefficient?

    a. No effect.

    b. Increase the value of the coefficient.

    c. Decrease the value of the coefficient.

    d. It would depend on the variance of the Y measures relative to the variance of the X measures.

    e. It would depend on the linearity of the regression.

7. The correlation between identical twins is .9 for height and .78 for weight. Jill is 63 inches tall and weighs 131 pounds. Both these measures are below average for the population. Decide whether the following statement is true or false: The best point estimates for the height and weight of Jill's identical twin sister, are 63 inches and 131 pounds respectively.

8. Explain the following analogy: Variance is to sum of squares as covariance is to sum of products.

9. What are the three fundamental assumptions underlying linear regression analysis?

10. Why would you expect the correlation between height and weight to be smaller for the population of professional basketball players than in the general population?

Short Problems

Problem 1 A university wishes to predict students' final year GPA from their GPA at the end of their first year. In order to formulate a prediction rule, the final year GPAs of a sample of 200 students are obtained along with the GPAs of these students at the end of their first year. For first year GPA, the mean was 2.2 and the standard deviation was 0.757. For the final year, the mean GPA was 2.6 and the standard deviation was 0.676. The product-moment correlation between the two GPAs for this sample was found to be r = .63.

a. What is the slope of the regression line for predicting final year GPA from GPA at the end of the first year?

b. A student obtained a GPA of 3.2 at the end of her first year. What is her predicted final year GPA?

Problem 2 An investigator sets out to test the hypothesis that the time it takes to perform a complex motor task decreases linearly as a function of the length of time in training. To evaluate this hypothesis, a total of 40 participants are assigned randomly to 4 conditions (10 in each) representing 5, 10, 15, and 20 hours of training. Set out the row headings of the linear regression summary table and enter the degrees of freedom for each.

Problem 3 Explain how the investigator in Problem 2 might go about evaluating whether the assumption of linear regression is appropriate (a) graphically and (b) by performing an analysis that obtains a mean square corresponding to "departure from linear regression." Set out the row headings of the analysis of variance summary table for the analysis along with the appropriate degrees of freedom.

Problem 4 Suppose the results from the analyses described in Problems 2 and 3 confirm the appropriateness of the linear regression model. The investigator reports that the fitted regression line shows that the time to complete the task drops from 40 minutes after five hours to 10 minutes after 20 hours.

a. What are the estimates of the parameters α and β of this regression model that the investigator obtained?

b. What is the predicted time to complete the task after 15 hours of training?

Problem 5 An investigator administers tests of spatial and numerical ability to a sample of 150 children. The standard deviations were 5.3 for the spatial ability test and 9.4 for the numerical ability test. The sum of products was 3736.0.

a. What is the value of the covariance between the two measures?

b. What is the value of the product-moment correlation coefficient between the two measures?

c. What is the regression coefficient for predicting spatial ability on the basis of numerical ability?

d. What is the regression coefficient for predicting numerical ability on the basis of spatial ability?

e. Explain in simple terms why these two regression coefficients are different.

Problem 6 An investigator administers a test of reasoning ability to a sample of 120 children ranging in age from 7 to 12 years. A regression analysis, regressing test scores on age, gives a value of 41.3 for $SS_{regression}$ and a value of 205.0 for SS_e.

a. What is the value of R^2 for these data?

b. What is the standard error of estimate?

c. Use an *F*-ratio to evaluate the significance of the linear relationship between age and reasoning ability.

Exercises in the Analysis of Realistic Data

Set 1: Understanding memory

Exercise 1. Short-term forgetting (Data Set MEM_01.dat)

In Exercise 1 of Chapter 1 you examined data from a short term memory experiment. The mean proportion of items correctly recalled is reproduced below. By examining the plot of these means decide on the plausibility of a linear regression model with retention interval as the predictor variable and proportion correct as the response variable.

RETENTION INTERVAL

	3 sec	6 sec	9 sec	12 sec	15 sec	18 sec
Mean	.80	.49	.35	.24	.13	.10

Set 2: Child Behavior

Exercise 2. Authoritarianism (Data Set CHILD_3.dat)

This study was described in Exercise 4 of Chapter 1 and the data for 65 mother-children pairs examined in Exercise 7 of Chapter 2. The results reported were as follows

	Mean	SD
Mother's authoritarianism	48.8	9.67
Child's school grade	68.2	7.40

A regression analysis treating children's school grade as the response variable and mothers' authoritarianism as the predictor variable gave a sum of squares for regression of 612.51 and a residual sum of squares of 2862.48. The covariance between the two measures is –29.90

a. Use these results to set out a complete regression summary table.

b. Calculate the product-moment correlation coefficient between academic achievement of the child and the degree of authoritarianism of the mother.

c. Calculate a 95% confidence interval for the regression coefficient.

d. Calculate the standard error of estimate.

e. Write out the regression equation for predicting children's school grade on the basis of mothers' authoritarianism score.

f. Is academic achievement of children related to the degree of authoritarianism of the mother, and if so in what way?

Exercise 3. Family routine chores and children's concern for others (Data Set CHILD_4.dat)

Is concern for others related to the performance of household chores? A study designed to explore this relationship was described in Exercise 8 of Chapter 2 along with some preliminary results as follows. The sample size was $N = 60$.

	Mean	SS
Family routine chores	23.8	4623.6
Concern for others	19.6	3634.8

A regression analysis treating children's concern for others as the response variable and time spent performing family routine chores as the predictor variable gave a sum of squares for regression of 841.76 and a residual sum of squares of 2792.98. The covariance between the two measures is 33.44. Use these results to:

a. Calculate the product-moment correlation coefficient between children's concern for others and time spent performing family routine chores

b. Calculate a 95% confidence interval for the regression coefficient.

c. Write out the regression equation for predicting children's concern for others on the basis of time spent performing family routine chores.

d. Is concern for others related to the performance of household chores, and if so in what way?

Exercise 4. Family ad hoc chores and children's concern for others (Data Set CHILD_8.dat.

Another category of chores is "family requested" consisting of family chores performed in response to ad hoc requests. Is concern for others related to the performance of this type of household chore? A study, similar to that described in the previous exercise and in Exercise 8 of Chapter 2, was conducted to answer this question. The only difference is that the response measure of time spent performing routine chores is replaced by time spent performing ad hoc chores. A new sample of $N = 50$ children was used.

A regression analysis treating children's concern for others as the response variable and time spent performing family ad hoc chores as the predictor variable gave a sum of squares for regression of 5.37 and a residual sum of squares of 1318.15. It also gives the sum of squares for response and predictor variables to be 1323.5 and 3043.4 respectively. The sum of products was $SP_{XY} = 127.88$. Use these results to:

a. Calculate the product-moment correlation coefficient between children's concern for others and time spent performing family ad hoc chores

b. Calculate a 95% confidence interval for the regression coefficient.

c. Is concern for others related to the performance of ad hoc household chores?

Set 2: Test Construction and Evaluation

Exercise 5. Test reliability (Data Set TEST_5.dat)

As the Director of Psychological Services for a large school board who has constructed a scholastic aptitude test, an important part of the evaluation of your test is to establish its reliability and validity. The reliability of a test refers to its stability; an unreliable test on which test scores for an individual would fluctuate from one occasion to another. One way of evaluating reliability is known as the test-retest method. In this method the test is administered to a sample and then at some later point in time re-administered to the same sample. Reliability is then measured as the correlation between scores on the two occasions. High reliability corresponds to high predictability from one test performance to another.

In order to obtain a test-retest measure of reliability the scholastic aptitude test is administered to a sample of 115 students, and then 3 month later re-administered to the same sample. Data Set TEST_5.dat contains the test scores for these two administrations. The standard deviations were 25.94 and 26.77 for the first and second administrations respectively. The sum of products was 66,817.13.

a. Describe what checks on the data you might perform before proceeding with the calculation of the correlation coefficient.

b. Calculate the covariance and the Pearson product-moment correlation coefficient between the two administrations.

Exercise 6. Test validity (Data Set TEST_6.dat)

Test validity is a general term referring to what it is that the test measures. Does your scholastic aptitude test really measure scholastic aptitude? You decide to evaluate the validity of the test by seeing how well it predicts scholastic performance. The scholastic aptitude test is administered to a sample of 110 students at the beginning of the academic year and at the end of the academic year the grade average of these same 110 students is recorded. Data Set TEST_6.dat contains the test scores and grade averages, expressed as percentages, of these 110 students.

Analysis of the data gave the following initial results. The means were 83.4 and 68.7 for the aptitude test scores and grade average respectively. The standard deviations were 28.4 and 9.8 for the aptitude test scores and grade average respectively. A regression analysis with aptitude test score as the predictor variable gave $SS_{regression}$ = 5434.9 and SS_e = 5036.9.

a. Describe what checks on the data you might perform before proceeding with the calculation of the correlation coefficient.

b. Set out the complete regression table for this analysis and calculate the relevant F-ratio.

c. Obtain the regression line for predicting grade on the basis of aptitude test score.

d. Obtain (i) the standard error of estimate, and (ii) a 95% confidence interval for the regression coefficient, for predicting grade on the basis of aptitude test score.

e. Calculate the Pearson product-moment correlation coefficient between aptitude test scores and grade average.

Exercise 7. Extroversion and aggression (Data Set TEST_7.dat)

The extroversion scale described in Exercise 10 of Chapter 2 was administered to a sample of 75 adolescent boys. A scale measuring aggression was also administered to the same sample. Is there a relationship between extroversion and aggression? The data from this study can be found in Data Set TEST_7.dat.

Analysis of the data gave the following initial results. The means were 59.2 and 33.5 for the extroversion and aggression scores respectively. The standard deviations were 16.6 and 11.1 for the extroversion and aggression scores respectively. A regression analysis with extroversion score as the predictor variable gave $SS_{regression}$ = 1907.5 and SS_e = 7271.2.

a. Use these results to set out a complete regression summary table.

b. Calculate the product-moment correlation coefficient between extroversion and aggression.

c. Calculate a 95% confidence interval for the regression coefficient.

10 CATEGORICAL RESPONSE VARIABLES AND DISTRIBUTION-FREE METHODS

Overview

1. Count data. Chapter 10 discusses data for which the response variable is categorical. In such cases the response measure consists of frequencies—a tally of the number of outcomes that fall into each of a defined set of categories. For this reason the data discussed in this chapter are often referred to as "frequency" or "count" data. One important difference between the statistical tests that were described in Chapter 10 and those of earlier chapters is that with count data assumptions about normal distributions and homogeneity of variance are no longer relevant. The basic strategy behind the analysis of count data is to evaluate models by comparing observed frequencies (f_o) with the expected frequencies (f_e) predicted by a model.

2. Goodness of fit. One class of models involves a single basis of classification. Testing models of this type is often referred to as evaluating goodness of fit. Residuals for models of count data are, as with other models, the differences between the observations and the predictions of the model. With count data this principle takes the form of residuals being the differences between observed and expected frequencies. The residuals can be evaluated using the Pearson χ^2 statistic. The chi-square distribution has one parameter, degrees of freedom. The χ^2 distribution is appropriate if the observed frequencies are independent and the expected frequencies not too small. The χ^2 procedure can be applied to goodness-of-fit situations with any number of categories. The degrees of freedom for χ^2 is $k - 1$ where k is the number of categories.

3. Contingency tables. A second class of models involves the analysis of data for which outcomes classified with respect to their values on two attributes. Such data can be arranged in a contingency table in which the categories of one attribute are represented in the rows and the categories of the second attribute are represented by the columns. The null hypothesis to be tested is that of independence between the row and column classifications.

4. Multiplication rule of independence. Independence is formally defined by the multiplication rule: The expected proportion in each cell is the product of that cell's marginal proportions. The multiplication rule is used to generate expected frequencies in each cell which are then used to calculate residuals. The residuals can then be evaluated using the Pearson χ^2 statistic.

5. Phi coefficient. For a 2×2 contingency table the phi coefficient (ϕ) provides a measure of the strength of the contingency. For larger tables a modified phi coefficient (ϕ_c, Cramér's phi) is used. The degrees of freedom for testing the null hypothesis of independence are $(R - 1) \times (C - 1)$ where R is the number of rows and C is the number of columns.

What You Should Know and Be Able to Do

■ Identify a response measure as categorical, and therefore yielding frequency (count) data.

■ Distinguish one-way (goodness of fit) designs from those intended to evaluate two-way contingency.

■ Formulate and apply the multiplication rule for independence of two events.

■ For contingency designs, arrange raw data in a two-way contingency table.

■ Formulate the appropriate null hypothesis for either one-way (goodness of fit) designs or those intended to evaluate independence in 2×2 and larger two-way contingency table.

■ Apply the formula for χ^2 and evaluate the result by using the table of the χ^2 distribution.

■ Calculate the phi coefficient (ϕ or ϕ_c) and describe its magnitude.

■ State the assumptions underlying the use of the chi square distribution.

Short-Answer Questions

1. What is the critical value of χ^2 in a goodness of fit test with 5 categories using $\alpha = .05$?

2. What is the critical value of χ^2 ($\alpha = .05$) you would use to evaluate contingency in a 3×4 contingency table.

3. Decide whether the following statements are true or false.

   a. In testing goodness of fit, the critical value of χ^2 does not depend on sample size.

   b. The value of χ^2 can be negative thus indicating a negative contingency, just as a negative correlation coefficient indicates a regression line with a negative slope.

   c. A zero contingency coefficient is an example of statistical independence.

   d. For a given significance level, the larger the degrees of freedom, the smaller is the critical value of χ^2.

   e. Other things being equal, the larger the value of χ^2, the better the goodness of fit.

4. In the formula for χ^2, what term appears in the denominator, f_o or f_e?

5. The use of χ^2 described in Chapter 10 relies on the assumption that the frequencies are independent. Explain what this means by giving an example for which this assumption holds, and one for which it does not.

6. Apart from the assumption of independence, what other feature of a contingency table might invalidate the use of the χ^2 test of independence?

Short Problems

Problem 1 For a certain lottery, it is known that 65% of winners are women and that 15% of winners are unemployed. If gender and employment status are independent, what is the proportion of unemployed women who win this lottery? Explain in words what this independence means.

Problem 2 In a study of brand preferences, 96 consumers are asked to select from three competing brands (A, B, and C) of the same product. The number choosing each brand is A:30, B:18, and C:48. Are these data strong enough to reject the hypothesis that the consumers show no brand preference?

Problem 3 In a follow up study of brand preferences, the investigator asks whether the pattern of brand preferences is the same for men and women. A total of 120 consumers (60 women, 60 men) are asked to select from three competing brands (A, B, and C) of the same product. For women, the number choosing each brand is: A:24, B:11, and C:25. For men, the number choosing each brand is A:16, B:27, and C:17. Do these data support the hypothesis of sex differences in the pattern of brand preference?

Problem 4 In a study of personal preferences, 200 subjects chose between two possible activities, A or B, and then chose whether they would prefer to engage in that activity in the day (D) or evening (E). The results showed that 150 chose A and 50 chose B, 80 chose day and 120 chose evening. Also 50 chose activity A in the day. Set out these data in a 2 × 2 contingency table, then evaluate the null hypothesis that choice of activity is independent of choice of time of day.

Problem 5 In order to test a mechanical card shuffling device, the machine performs a total of 100 trials (shuffles). Each trial consists of shuffling the deck, drawing one card, and recording the card's suit: clubs, hearts, spades, or diamonds. The card is then returned to the deck and the deck reshuffled ready for the next trial. The frequency with which each suit was selected over the 100 trials is shown below.

Clubs	Hearts	Spades	Diamonds
21	31	29	19

If the machine is shuffling properly then, for any trial, each of the four suits has an equal chance of being chosen. Calculate the value of the test statistic you would use to decide whether these results support the claim that the machine is shuffling properly.

Problem 6 In an experimental evaluation of a program designed to help people stop smoking, half the members of a sample of 72 smokers are randomly assigned to an experimental group, the other half to a control group. After 2-months, 23 members of the experimental group and 12 members of the control group have stopped smoking. Do the data support the claim that the experimental procedure was effective?

Problem 7 A sample made up of 65 registered Republicans and 75 registered Democrats contains 85 people who favor an affirmative action program and 55 who are opposed. Of the 85 who favor the program, 48 are Democrats. Set out these data in a 2×2 contingency table, calculate a contingency coefficient between party affiliation and opinion, then evaluate whether the contingency is statistically significant.

Problem 8 A survey asked 20 women and 20 men whether they would vote yes (Y, y) or no (N, n) if asked if they approved of capital punishment. The responses of women were coded using upper-case letters, men with lower-case letters. The 40 observations were as follows.

N y N Y n y N y Y N N n y N Y n N N n y Y y y N n y Y N y n Y y y N Y y y N y

Arrange these responses in a 2×2 contingency table, then calculate a contingency coefficient and evaluate whether the contingency is statistically significant.

Problem 9 Members of a sample of 84 people receiving psychotherapy are assessed for anxiety and depression. It is found that 51 are classified as high-anxious, 33 as having a normal anxiety level. The depression scale classifies 40 as depressive and 44 as nondepressive. It is also found that 23 of the 84 people are neither depressed nor anxious. Set out these data in a 2×2 contingency table, calculate a contingency coefficient between anxiety and depression, then evaluate whether the contingency is statistically significant.

Exercises in the Analysis of Realistic Data

Exercise 1. Age and alcohol-related fatal driving accidents

This study examines the relationship between age and alcohol-related fatal driving accidents. Of the 200 fatalities examined, 26 involved drivers under 21 years of age, 95 involved drivers 21-34 years of age, 44 involved drivers 35-44 years of age, and 35 involved drivers older than 44. The percentage of licensed drivers in each of these four age ranges is 7%, 30%, 22% and 41% respectively. Use these percentages to derive the expected number of fatalities for each age group for a sample of 200 under the hypothesis that fatality rates simply reflect the percentage of licensed drivers in each age group. Evaluate whether the observed frequencies are compatible with these expected values.

Exercise 2. Imitating aggression

A sample of 92 participants complete an aggression assessment scale before and after seeing a movie. Half of these 92 participants saw a movie with a great deal of violence, the other saw a travel film with no violence. Of the participants who saw the violent movie, 31 displayed increased levels of felt aggression, whereas 11 of the participants who saw the travel movie displayed increased levels of felt aggression.

  a. Using $\alpha = .05$ (two-tailed), what is the tabled critical value of the test statistic you would use to evaluate the null hypothesis that the movie has no influence on felt aggression?

  b. Based on the above results, calculate the value of the test statistic you would use to evaluate the null hypothesis that there is no relationship between type of movie and increase in felt aggression, and decide whether the null hypothesis should be accepted or rejected.

  c. Calculate a contingency coefficient for these data.

Exercise 3. Sources of persuasion

Are arguments more or less persuasive depending on the source (as opposed to the strength) of the argument? To answer this question an investigator presented volunteer college students with arguments in favor of requiring students to pass a set of comprehensive exams in order to graduate. Thirty volunteers in one group (Group A) were told that the arguments had been prepared by a Commission on Higher Education; 30 volunteers in another group (Group B) heard the same arguments but were told they had been prepared by a high school student. After hearing the arguments the volunteers recorded whether they agreed or disagreed with the proposal. In Group A, 21 agreed and 9 disagreed; in Group B, 12 agreed and 18 disagreed. Use these results to address the claim that arguments are more persuasive depending on the source of the argument.

Exercise 4. Daycare and child security

In a study of the relationship between an infant's sense of security and daycare, Barglow, Vaughn, and Molitor (1987) classified 110 infants as either Secure (S), Insecure-avoidant (IA), or Insecure-resistant (IR). Fifty-six of the mothers cared for their infants at home and 54 worked outside the home and used substitute care in the home. The resulting 2×3 contingency table is as follows

	S	IA	IR
Mother at home	40	5	11
Mother at work	29	17	8

Use these data to evaluate the claim that there is a relationship between the infant's security and the nature of daycare. Would you consider the two categorical variables used in this study as natural or manipulated? What is the relevance of this distinction to the interpretation of the findings of this study?

Introduction to the Use of SPSS for Windows

1   Introduction to SPSS: Reading in Data from Disk

2   Examining and Describing Data

3   Obtaining Standard Errors and Confidence Intervals

4   Analysis of Variance for Independent Groups

5   Matched Pairs and Within-Subjects Designs

6   Regression and Correlation

7   Analysis of Contingency Tables

1 INTRODUCTION TO SPSS: READING IN DATA FROM DISK

What Is SPSS?

SPSS is short for Statistical Package for the Social Sciences. From its early introduction in the 1980s it has become one of the most popular software packages among research psychologists. The software has been translated into numerous platforms, running various operating systems including: main frame computers running UNIX or VMS, IBM PC compatibles running DOS, Windows, Windows '95, and OS2, and Macintosh. Traditionally, one used SPSS by writing short computer programs telling SPSS which modules it should load, and what it should do with your data. In its most recent releases, however, SPSS has been redesigned to take advantage of the windowing environments of modern operating systems. The two most likely packages you will encounter are version 7 of SPSS for Windows (which also runs on Windows 95) or the Student Edition of SPSS, which is very similar. All of the examples in this workbook were generated and tested using these editions. Accordingly, there are no listings of the older SPSS command codes. Should you be working with a pre-windows version of SPSS, consult any of the many specialty books written on the subject.

Additionally, this workbook cannot provide instructions on how to operate the Windows environment. We will assume that you know the basics—of how to turn on your computer, find files in directories or folders, etc. If this is the first time you've used a computer you should avail yourself of whatever help your course instructor or teaching assistants can provide.

The conventions we use throughout this workbook are as follows. Any time we want you to select a menu or menu item from SPSS we print its name in **boldface**. Additionally, some menus and menu items have keyboard short cuts available. This is usually is apparent in the name of the item as the first (or sometimes a later) letter of the name underlined, for example **File** which is the menu containing most of the basic commands relating to working with files on your computer. We always use the name with the underline, in order to avoid confusion between the format of this book and the format of your screen. To use one of these short cuts you hold down the **Alt** key on your keyboard while simultaneously pressing the underlined letter. When we instruct you to do this, we abbreviate this to **Alt-F**. Similarly, if we wants you to hold down the Control Key (usually labeled Ctrl on your keyboard) while pressing the F key, we instruct you to press **Ctrl-F**.

In addition to using keyboard commands, a Windows program like SPSS is really at its best when you are using a mouse or other pointing device (like a track ball or touch pad). When we want you to select some option using your mouse we will tell you to click (or double click as appropriate) on something. This means to use your pointing device to move the cursor on top of the object we mentioned and then to press the selection button once (or twice if we tell you to double click).

Whenever we wish to present some output from an SPSS job, we enclose the output file in a box, like this one. We make every effort to ensure that the output we present looks like the output you will see on your screen. In the interest of saving pages, however, we occasionally shrink charts and graphs by up to 50%.

Often the program responds to an action on your part by producing a new small window, called a dialogue, which has other options (buttons you can press, places you can type in information) for you to choose from, based upon the initial action you performed. We tell you what settings to use in any given dialogue throughout the examples in the workbook, but you should always feel free to explore the results of other actions. Like exploring any new environment, trial and error navigation often leads to the most familiarity with the new area. Always feel free to try things out. Your computer and SPSS are much more sturdy and forgiving than you fear they are. You can almost always undo anything you've done that you don't like. Finally, there are a few differences between the full version of SPSS and the student edition. [When we wish to highlight one of these differences, we provide the instructions for the student version in square brackets, like these.]

Before we can begin with any statistical analysis we will get some data into the system for SPSS to work with. There are two basic methods we can use to have data available for SPSS. We can enter it directly in the Data Editor or we can have SPSS read it from a text file.

Entering Data from the Keyboard: The Data Editor

The Data Editor is one of the windows (perhaps the only one) which will open when you first start SPSS for windows. [In the student version, this window is labeled Newdata.] If you look at the contents of the window it looks like a paper spreadsheet with rows, each labeled with ascending numbers and columns, each labeled var.

Example 1

	var	var	var	var
1				
2				
3				
4				

To begin entering data, you double-click on one of the var boxes to label that column with a variable name. When you double-click on any of the var boxes a new dialogue will open called Define Variable. In the top of this dialogue is a white text box labeled Variable Name into which you can type the name you wish to give this variable. Choosing Data Set 2.4 from Section 2.1 of the text book we might label this first variable Age or Group, since we will use this column to tell SPSS whether the data for this observation came from one of the children in the six-year-old group or one of the children in the eight-year-old group. A variable name can be any string of up to eight letters long. At the bottom of this dialogue are four buttons labeled: **Type**, **Labels**, **Missing Values**... and **Column Formats**... . Pressing any of these buttons will call up additional dialogues to help you further define your newly named variable. We will discuss only the **Type** button here and will save the **Labels** button for Section 2 of the workbook.

Pressing (clicking on) the **Type** button will call up a dialogue called Define Variable Type. It is in this dialogue that you tell SPSS what type of data will be stored under this variable name. You select a type by pressing one of the radio buttons next to a type name. Only two of these types need concern us for now.

The first is numeric, which is the default setting for any variable. Setting the type to numeric means that the values of this variable will only be numbers. In addition to choosing numeric you can further format the numbers that will be in this column by entering values in the **Width** and **Decimal Places** text boxes at the right of the dialogue. The default setting for these are 8 and 2 respectively, which means that numbers in this variable can be up to 8 characters wide, of which two can be displayed to the right of a decimal place. These settings should be adequate for any examples in the text or work books, but feel free to experiment with different values, or even different types, at your convenience.

The second important type is string, which is what you should select if any of the values for that variable will contain letters, rather than numbers. For example, if you had a variable called Gender in your data set, which could take values of F for female or M for male, you should set the type of the variable Gender to string.

Once you have set up all the variables for your data set you can begin entering values for them in the columns below, making sure that you keep the values from any one observation on the same row. Thus to continue with Data Set 2.4 from Section 2.1 of the text, we might begin entering:

Example 2

	Age	Vocab	var	var
1	1.00	.75		
2	2.00	.76		
3	1.00	.71		
4	2.00	.66		

Note that although we entered a value of 1 for the first age group (six-year-olds), SPSS automatically changed it to 1.00 in order to bring it into line with the Decimal Places set under the numeric data type. If we wanted it to read 1 instead, we should have set the Decimal Places to 0 in the Define Variable Type dialogue. Also note that although we have chosen to alternate data from six-year-olds and eight-year-olds, to demonstrate the full range of the data set in the example, there is no reason why you should feel compelled to do likewise. In fact, it would probably prove faster to type in all the data from the six-year-olds first and then all the data from the eight-year-olds, since that is how the data set is laid out in the text book.

On a final note, we chose this data file for a reason, namely that it is quite lengthy. Obviously it would be to our advantage if we didn't have to type it all out ourselves. SPSS can help us by reading the data in directly from a text file.

Reading Data from Files

You should feel free to type in any data from your text book or work book to try them out; however there is an easier way of handling the data from the examples in the textbook and workbook. All of the data sets used in either book are available as text (ASCII) files either from your instructor, or by downloading the files from the book's web site.

Data Sets from the textbook have file names corresponding to their designations in the textbook itself. Thus Data Set 2.1 in the textbook adopts the file name **DS02_01.dat**; Data Set 2.4 has the file name **DS02_04.dat**; data Set 3.1 has the file name **DS03_01.dat**, and so forth. The leading zeros (01, 02, etc.) have been inserted so that when the files are listed on your screen they will be correctly ordered. Without the leading zeros, most software when reading filenames will strictly alphabetize by placing the number 11 before the number 2. However 02 will be placed before 11. The filename extension ".dat" has been added because both SPSS and SAS look for this extension as designating a data set. Data Sets from the Workbook have file names denoting one of the three research areas used in the Workbook Exercises. Data sets for the memory area are named **MEM_01.dat**, **MEM_02.dat**, etc. Data sets for the child behavior area are named **CHILD_1.dat**, **CHILD_2.dat**, etc. Data sets for the exercises involving test construction and evaluation are named **TEST_1.dat**, **TEST_2.dat**, etc. For convenient reference, all data sets are listed in Appendix B of the workbook.

To use one of these data files, you will first have to copy it to a directory on the computer on which you intend to use SPSS. Ask your instructor if you are uncertain about how to do this. For now, we will assume you have copied all of data files to a directory on your computer called **datasets**.

To read in a data file from this directory first select the **File** menu and then choose the menu item **Read ASCII Data**. This will bring up a tiny menu asking if you want to read the data **Freefield** or in **Fixed Columns**. You should choose Freefield as that will work for all the examples in this book. Freefield causes SPSS to read across a row of data in the file, looking for one value for each variable, separated by spaces, commas, or whatever. Fixed Columns on the other hand requires you to know (or to count) how many columns wide each variable is, and how many columns are wasted between the values of different variables. In general Freefield will be easier. We aren't done yet. Now that we've selected Freefield we will have to tell SPSS how many variables it is looking for, and in what file it should look for them. We do this in the Define Freefield Variables dialogue which now opens.

At the top of the dialogue is a section labeled File, with a button labeled Browse. When you click on this button, you will be taken to a display of your hard drive's contents which should be somewhat familiar if you've used Windows or Windows 95 before. Essentially, use your mouse to select the appropriate directory (we are using c:\datasets as we suggested) and then select the file you wish to open (we select **DS02_04.dat**).

[In the student edition, the process is essentially similar, though differing in detail. When you first select **Read ASCII Data** you will be sent to a dialogue which will contain both the buttons for choosing Freefield or Fixed format, as well as the windows boxes for choosing the specific file to load. Thus steps one and two of the above method are combined in the student edition. Having selected Freefield and the file of your choice you then press the **Define** button, and proceed to the define variables dialogue as below, with the name of your file already listed in the upper left, and no opportunity to Browse to a different file. You proceed, defining each variable in turn, as below.]

Next we have a text box labeled name. Here you type in the name of the first variable in the file. (Although you are free to use your own label, the table listing data sets in the appendix of the workbook provides suggested labels that can be used for this purpose.) Immediately below this text box are a pair of buttons labeled numeric and string, which you use to set the type of the variable you just named. Additionally, if you select numeric you can provide values for the width and decimal places just as you did when typing in the data manually. Once you have defined the first variable (which we had called Age when doing this example in the Data Editor) you press the button labeled **Add** the name of the variable you just defined will appear in the list box next to the button, and the text box near the top of the dialogue will be blanked, ready for you to name the next variable (In our example Vocab). After you have added all the variables you are loading from the data file, you then press the **OK** button at the top of the dialogue, and you're finished. All the data, and variable names should now appear in the Data Editor window, exactly as if you had typed them in by hand. You can now double click on any of the variable names you wish, to make any adjustments in name or type, from the Define Variable dialogue, exactly as we did above.

Saving Your Work

Regardless of whether you entered the data for this example manually using the Data Editor or had SPSS read the data directly from a file, now that you have entered it, you probably should save the data so that you can use it at a later time. To save a file you can select the **File** menu and then choose the **Save** menu item, or you can press **Ctrl-S**. In either case a save dialogue will appear in which you can select which directory you want to save the data in (the default will be whatever directory SPSS has worked with most recently) and what you wish to call the saved data file. You can select any file name you wish, up to eight characters (or longer in Windows 95), and SPSS will automatically append the extension .sav for saved file. To finish saving, and exit the dialogue press the **Save** button. We saved the data file under the name **DS02_04.sav**. The .sav extension will distinguish this saved file from the original ASCII file, and at the same time the label preserve, the correspondence between the two files.

Reading in Previously Saved Data Sets

If the data have been previously saved as a .sav file they can be read in directly. From the **File** menu select **Open** then **Data**. This brings up the Open Data File dialogue. Locate the drive, folder, and file you want, then click OK.

Transforming Data and Creating New Variables

Once you have the data entered you can use some basic functions to create new variables or transform existing ones. By selecting the **Transform** menu, and then choosing the **Compute**... item, you can create new variables based upon arithmetic operators and mathematical functions. Making these selections opens the Compute Variable dialogue. In this dialogue you see a small text box in the upper left, labeled target variable. Here you can type the name of a new variable you wish to create. As you can see the new variable is said to be equal to the larger, blank text box to the right. In this text box you can insert any of your current variable names joined by the arithmetic operators listed below, and/or set any of them to be the argument of one of the mathematical functions listed in the scrolling list box to the right of the dialogue. For example, data for the vocabulary scores are expressed as proportions. Suppose you wished to transform the proportions into percentages by multiplying each score by 100. To make this transformation with our data we would open the Compute Variables dialogue, and in the left text box enter a name for our transformed variable, say transvoc, since it is the transformation of our vocab variable. In the right text box we would type the desired transformation, in this example vocab*100. If you press the OK button, the dialogue will close and we will be returned to the Data Editor window where the new variable will appear next to our old variables.

Try experimenting with other types of transformations. How, for example would you create a variable that was the average of two variables already in your data set? Don't forget to save your data set again, after you've created any new variables, or they'll be lost.

Listing Your Data Set

Finally, having entered your data, and gotten it into a format you like, with any new variables you have created you may want to take a look at the finished product. You can obtain a listing of a data file by selecting the **Print** item from the **File** menu, or by pressing the **Print** button on your toolbar (the one that looks like a picture of a printer).

Review of Concepts

■ Using SPSS to analyze your data involves using windows commands to select appropriate menu choices, and refine those choices using dialogues.

■ Data can be entered into SPSS in one of two ways. It can be entered manually using the data editor, or it can be read from a disk file. [The number and order of steps followed to read data from a file differs between the full and student editions of SPSS. See the section Reading Data from the Files for details.]

■ Regardless of the method of entry you can define your variables, including naming them and defining their type. Two types are string variables and numeric variables. Numeric variables can be further defined in terms of their width and decimal places.

■ The PRINT menu item or button lists our data set. The Save item saves it.

■ The Compute Variables dialogue from the Transform menu can be used to create new variables or transform existing variables using arithmetic operators and mathematical functions.

Exercises

The best way to learn any new procedural skill is by trial and error. We therefore encourage you to try loading some data sets, saving the file, and then listing the results. Don't worry, as long as you use the tools we've shown you in this chapter and as long as you let SPSS choose the extensions for any files you save, you can't permanently damage any of the data sets on your disk.

The exercises at the end of this and subsequent sections are intended to get you through the initial stages of becoming familiar with the use of SPSS to perform the procedures described in that section. There are many more data sets from the textbook and the workbook on which you can practice.

As a starting point, enter Data Sets 2.1 through 2.4 either from the keyboard or from the ASCII files **DS02_01.dat**, **DS02_02.dat**, etc. These files will be used in the next section. Suggested variable names will be found in the listing of data sets in Appendix B of this workbook. Note that Data Sets 2.1 and 2.2 are non-numerical and the variables must be declared "string." Save each of these files either with a name of your own choice, or preferably simply as DS02_01.sav, DS02_02.sav, etc. List the files and check that the data were entered correctly. Continue entering data until you are thoroughly familiar with the procedure. If you save your files, none of this effort will be wasted because, when performing later exercises, you will be able to call up the data directly from the saved (.sav) file.

EXAMINING AND DESCRIBING DATA

Frequencies

Chapter 2 of the textbook begins with the notion of frequencies and methods for tabulating and graphically representing them. In SPSS, most of these operations are carried out by selecting **Summaries** from the **Statistics** window and then pressing **Frequencies**.

Follow this sequence using the data file **DS02_01.dat**. If you select the default options and label the sole variable "Response", will produce the following output. These data correspond to the first example in Chapter 2 of the textbook (see page 107). The bar chart was not one of the default options, but was obtained by pressing the **Chart** button on the Frequencies dialogue. Performing the same sequence of operations while using the data file **DS02_02.dat** and labeling the variable "Brand" will produce the results shown on page 108. The chart was produced by selecting as above but choosing percent rather than frequencies when selecting the chart.

Note that we have obtained percentages rather than proportions as used in the second example in the textbook. This is because SPSS does not give proportions as an option. However, the results are functionally equivalent because percentages are just proportions rescaled to be out of 100 rather than 1.

To examine frequencies of continuous data, we normally use a histogram, rather than a bar chart. The example on page 109 is taken from **DS02_03.dat**, with the variable named IQ.

The histogram was obtained by selecting histogram from the chart options in our usual frequencies dialogue. This is our first opportunity to exercise this option because we now have quantitative data. Note that this output has suppressed the frequency table that would normally be part of this output. This suppression is achieved by deselecting the "Display frequency tables" check box. This histogram uses class intervals of width 5 as does Figure 2.5 in the text. However, the two graphs are not identical because they make their cuts at different points. Figure 2.5 begins the first interval at 65, resulting in intervals 65–69, 70–74, 75–79, etc. whereas the SPSS histogram begins the first interval at 68, resulting in intervals of 68–72, 73–77, 78–82, etc. Notice also that the figure records the midpoint of the interval rather than the interval boundaries.

It is important to realize that both these forms of the histogram are equally acceptable and that it is not a matter of one being right and the other wrong. Both are accurate portrayals of the data. Recall that histograms serve the purpose of displaying the data in order to detect the presence of features such as skew and outliers. Either form of the graph will serve this function.

We do, however, have the option of modifying charts to suit our preferences. To exercise this option simply double click on the chart as it appears in the Output Navigator window, and after a moment or two, a new window will open labeled "SPSS Chart Editor". In this window you can modify virtually any aspect of your charts. By selecting the **Chart** menu and then the **Axis** control you can set up your class intervals. Simply select Interval in the Axis Selection dialog box. Under Intervals, select Custom and then Define. By choosing an interval width of 10 and resetting the axis minimum and maximum to 60 and 150 respectively, you can recreate Figure 2.4 from the textbook.

Frequencies

Statistics

	N	
	Valid	Missing
RESPONSE	26	0

RESPONSE

		Frequency	Percent	Valid Percent	Cumulative Percent
Valid	FM	3	11.5	11.5	11.5
	HO	6	23.1	23.1	34.6
	LP	4	15.4	15.4	50.0
	MO	5	19.2	19.2	69.2
	TP	8	30.8	30.8	100.0
	Total	26	100.0	100.0	
Total		26	100.0		

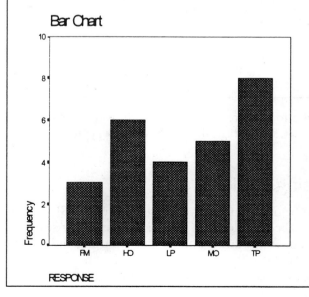

Bar Chart

Frequencies

Statistics

	N	
	Valid	Missing
BRAND	50	0

BRAND

		Frequency	Percent	Valid Percent	Cumulative Percent
Valid	A	6	12.0	12.0	12.0
	B	11	22.0	22.0	34.0
	C	18	36.0	36.0	70.0
	D	10	20.0	20.0	90.0
	E	2	4.0	4.0	94.0
	F	3	6.0	6.0	100.0
	Total	50	100.0	100.0	
Total		50	100.0		

Bar Chart

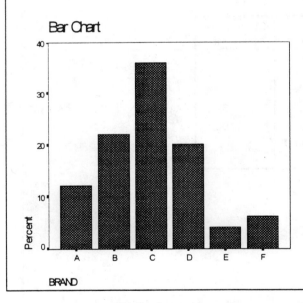

To further reduce the amount of effort, there is even a shortcut to this process. By simply clicking on the chart near the axis, when its in the editor window, you don't need to use the menus.

Using Explore

One of the most useful places to begin examining your data is with the SPSS built in exploratory data analysis engine, called Explore. It can be selected after selecting **Summarize** from the **Statistics** menu. Activating **Explore** and choosing the default options for our variable IQ will produce the following output. Take a moment to examine it on pages 110–111 or try it yourself using the data file **DS02_03.dat**.

Frequencies

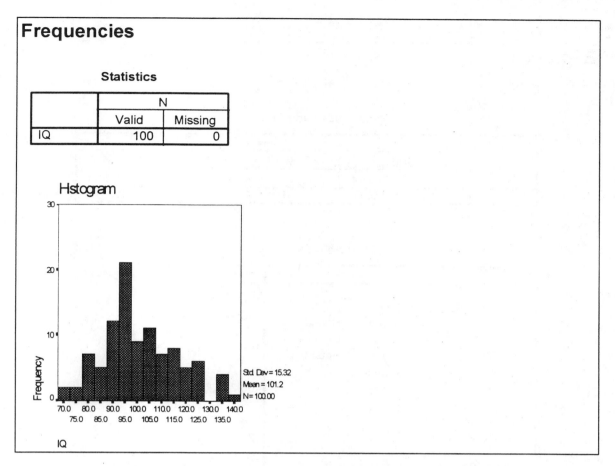

Statistics

	N	
	Valid	Missing
IQ	100	0

Histogram

IQ

Std. Dev = 15.32
Mean = 101.2
N = 100.00

The first section lists the variable name and the number of cases processed. It is always a good idea to double check these values whenever you get output from Explore or any other SPSS module. This way you know that you and SPSS are talking about the same variable, and that you agree about how many subjects there are in your sample. (Note that SPSS uses upper-case N to refer to sample size.) If the N looks wrong, it might mean that SPSS had trouble reading your data and you may want to go back and look at your data file for errors.

Next you are presented with a table labeled Descriptives, which presents many of the numerical descriptors of your data with which you should already be familiar. The output indicates that for this data set (**DS02_03**) we have an N of 100, a Mean of 101.17, a Standard Deviation of 15.32, and a Variance of 234.84. One other value worth mentioning in passing is Skewness of 0.435. You were introduced to skew as a visual cue you could look for in plots of your data. The results presented here is simply a numerical summary calculated directly from the data. A large number for Skewness reflects a problem which should also be apparent on visual examination, negative and positive values of Skewness reflect negative and positive skewing, respectively. None of the other listed values listed need concern you. They all have limited use, for quite specific applications.

Additionally, this section lists the median (98.00), the minimum value, the maximum value, the overall range of the data, and the interquartile range. Additional descriptive statistics can be requested from the Explore dialogue.

This section is followed by three plots, a histogram like we produced previously, a stem-and-leaf display and a box-plot. Each of these is a separate graphing object and can be modified by double clicking on them in the output navigator. Note that the

Explore

Case Processing Summary

	Cases					
	Valid		Missing		Total	
	N	Percent	N	Percent	N	Percent
IQ	100	100.0%	0	.0%	100	100.0%

Descriptives

		Statistic	Std. Error
IQ	Mean	101.1700	1.5325
	95% Confidence Interval for Mean Lower Bound	98.1292	
	Upper Bound	104.2108	
	5% Trimmed Mean	100.7222	
	Median	98.0000	
	Variance	234.850	
	Std. Deviation	15.3248	
	Minimum	69.00	
	Maximum	142.00	
	Range	73.00	
	Interquartile Range	19.0000	
	Skewness	.435	.241
	Kurtosis	-.007	.478

stem-and-leaf display separates off outliers or extreme values (in this case only the value 142 is extreme), as does the box-plot which prints a 0 for values beyond the whiskers.

Descriptives by Groups

Occasionally we have natural groups of subjects in our data set and would like to be able to obtain descriptive statistics separately for each group. In Chapter 2 of the textbook, we saw data consisting of scores on a vocabulary test for two groups of children, six-year-olds and eight-year-olds. This data set can be found in the file **DS02_04.dat**. The data set contains two numerical variables, the first refers to the predictor variable and should be labeled group. Its values consist of a 1 if the child is a six-year-old and a 2 if the child is an eight-year-old. The second column is response variable—the score on the vocabulary test.

IQ

```
        IQ Stem-and-Leaf Plot

 Frequency      Stem &   Leaf

      1.00        6 .   9
      2.00        7 .   14
      3.00        7 .   599
      8.00        8 .   00122344
      7.00        8 .   5788889
     12.00        9 .   111222233333
     18.00        9 .   556666666677777788
     14.00       10 .   00011223334444
      9.00       10 .   566788899
      9.00       11 .   113334444
      3.00       11 .   789
      8.00       12 .   01233344
      1.00       12 .   7
      1.00       13 .   3
      3.00       13 .   577
      1.00   Extremes    (>=142)

 Stem width:     10.00
 Each leaf:       1 case(s)
```

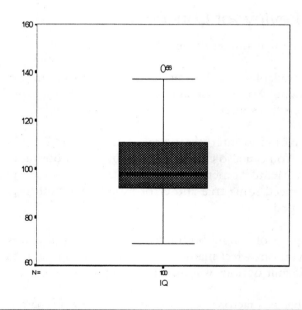

In order to make our analyses more intelligible, it would be helpful to have the groups labeled as something more memorable than 1 and 2. Select **Define** variable from the **Data** menu. When the dialogue opens, select the variable group (or whatever you've called it) and create Value Labels, in the area below. We chose to create the Value Labels "six-year-olds" if Group = 1 and "eight-year-olds" if Group = 2. Remember to press the Add button after creating each label, they should then be displayed in a list on the dialogue. Creating value labels for groups in your data is usually a good idea, as researchers often return to their data after long periods of time, and may not remember whether males were 0 or 1 in this particular set, for example. Meaningful Value Labels also make your graphs and tables much more informative.

Having labeled our data, we can now proceed with the descriptive statistics. For this example we will use the **Explore** option, as above. The only difference is that when we select the variable Vocabulary for descriptives, in the Explore dialogue, we will also select Group, and move it to the field labeled Factors. In addition to the default statistics and plots, as above, we have selected Histograms as an additional plot from the Plots control. As a result, we receive the output shown on pages 113–115.

Note that the default for using explore with a grouping variable as a factor is to present all the statistics and plots separately for each group, except for the box-plots which are presented side-by-side to allow direct comparison, as in Figure 2.26 of the text book. Unfortunately, there is no option provided to select back-to-back stemplots as in Figure 2.20 of the text. This is a limitation common to all statistical software packages, at the present date.

In conclusion then, SPSS provides many tools for examining both numerical and graphical descriptions of your data. The most useful of these, and the easiest to apply, is the Explore option. Separate descriptions can be obtained for groups in your data by selecting the grouping variable and using it as a factor in Explore. Finally, value labels will make such a grouping variable more intelligible to your audience, and to yourself, as time advances.

Review of Concepts

- SPSS can be used to obtain many numerical and graphical descriptions of your data.

- Frequency statistics are available by selecting Frequencies from the Summarize option of the Statistics menu. You can obtain tables of frequencies as well as bar charts and histograms from this source.

- You can vary the number of class intervals in your bar charts and histograms by editing the graph directly. You can also create class variables by choosing to Recode your data from the Data Menu. Sometimes recoded data will not be plotted as you'd expected, you may need some trial and error experimentation to get things looking the way you'd hoped.

- Explore is a powerful tool for obtaining both numerical and graphical descriptions of your data. In explore you can select many different statistics and graphs for presentation, but even the default options will usually be useful.

- By naming a group variable as a factor in Explore you can obtain separate numerical and graphical descriptions for each group in your data. You can also perform side-by-side comparisons of the groups with the box-plots produced. Always remember to provide informative Value Labels for your group variable(s) by selecting Define Variable from the Data menu.

Explore
GROUP

Case Processing Summary

	GROUP	Cases					
		Valid		Missing		Total	
		N	Percent	N	Percent	N	Percent
VOCAB	six-year-olds	50	100.0%	0	.0%	50	100.0%
	eight-year-olds	50	100.0%	0	.0%	50	100.0%

Descriptives

	GROUP			Statistic	Std. Error
VOCAB	six-year-olds	Mean		.6010	2.4E-02
		95% Confidence Interval for Mean	Lower Bound	.5534	
			Upper Bound	.6486	
		5% Trimmed Mean		.6002	
		Median		.5850	
		Variance		2.8E-02	
		Std. Deviation		.1676	
		Minimum		.27	
		Maximum		.95	
		Range		.68	
		Interquartile Range		.1850	
		Skewness		.113	.337
		Kurtosis		-.299	.662
	eight-year-olds	Mean		.7672	2.1E-02
		95% Confidence Interval for Mean	Lower Bound	.7258	
			Upper Bound	.8086	
		5% Trimmed Mean		.7743	
		Median		.8000	
		Variance		2.1E-02	
		Std. Deviation		.1458	
		Minimum		.42	
		Maximum		.98	
		Range		.56	
		Interquartile Range		.2025	
		Skewness		-.728	.337
		Kurtosis		.016	.662

VOCAB
Histograms

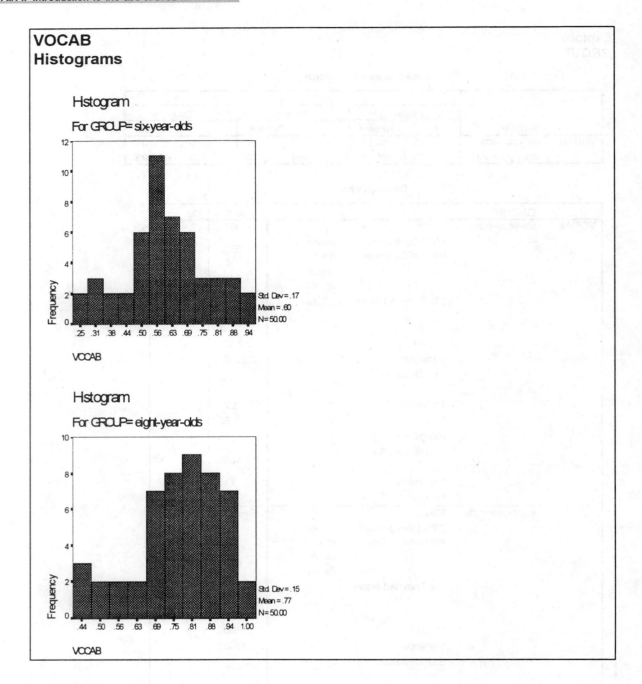

Histogram

For GROUP= six-year-olds

Std. Dev = .17
Mean = .60
N = 50.00

VOCAB

Histogram

For GROUP= eight-year-olds

Std. Dev = .15
Mean = .77
N = 50.00

VOCAB

Stem-and-Leaf Plots

```
VOCAB Stem-and-Leaf Plot for
GROUP= six-year-olds

 Frequency    Stem &   Leaf

      2.00      2 .  78
      4.00      3 .  3446
      5.00      4 .  02389
     15.00      5 .  112344555677889
     12.00      6 .  112233367799
      5.00      7 .  12589
      5.00      8 .  23889
      2.00      9 .  35

 Stem width:       .10
 Each leaf:       1 case(s)

VOCAB Stem-and-Leaf Plot for
GROUP= eight-year-olds

 Frequency    Stem &   Leaf

      2.00      4 .  23
      2.00      4 .  58
      1.00      5 .  3
      2.00      5 .  79
      2.00      6 .  34
      5.00      6 .  66779
      4.00      7 .  0144
      6.00      7 .  556788
      9.00      8 .  000011233
      7.00      8 .  5566788
      5.00      9 .  01344
      5.00      9 .  55688

 Stem width:       .10
 Each leaf:       1 case(s)
```

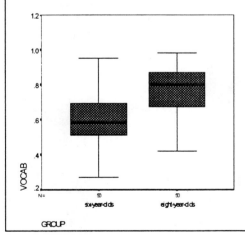

Exercises

1. Enter Data Set 2.5 from the keyboard or from the file **DS02_05.dat**, noting that this is a "string" variable. Obtain a frequency distribution of the data. Remember that suggested variable names can be found in the listing of Data Sets in the Appendix of this Workbook.

2. Look at Problem 2 in Chapter 2.2 of the text book. The data for this problem can be found in the file **DS02_06.dat**. Obtain descriptive statistics for the sample as a whole, and separately for each group. How do the groups compare to one another, numerically and graphically?

3. Using data in the file **DS02_07.dat** from Problem 3 of Section 2.1, obtain a stemplot, the mean, median, and standard deviation of the data.

4. Using data in the file **DS02_08.dat**, check your answers to Problem 5 of Section 2.2.

5. Using data in the file **DS02_09.dat**, obtain the variance of each data set used in Problem 8 of Section 2.2.

6. Using data in the file **DS02_10.dat**, obtain the standard deviations of each data set used in Problem 9 of Section 2.2.

3 OBTAINING STANDARD ERRORS AND CONFIDENCE INTERVALS

In Section 2 of this workbook we left off our discussion of descriptive statistics having obtained separate descriptions for two different groups in our data set, as well as a graphical comparison of the groups using side-by-side box-plots. Section 3 explores methods in SPSS for determining whether two groups really do differ. Students who feel the need to review the notion of group differences should consult chapters 5 and 6 of the text book.

The example we will be working with is described in Section 6.2 of the text. It consists of data from an experiment on the effectiveness of desensitization therapy for the reduction of anxiety in patients with phobias. The data can be found in file **DS03_01.dat**. The first column consists of the group information, with a 1 indicating that that subject was part of the control group in this study, and a 2 indicating that they received the desensitization therapy. The second column is a measure of anxiety, the recorded increase in the patient's heart rate upon presentation of the phobic stimulus, after completion of the therapy (or control). We loaded the data into SPSS as described in Section 1, calling the first variable Group and the second Anxiety.

In order to facilitate the interpretation of these data, we provided some informative labels for the variables. From the **Data** menu we selected **Define Variables,** and then pressed the Labels button on the resulting dialogue. For the Group variable we supplied the Variable Label "Therapy" and the Value Labels "control" if group = 1 and "desensitization" if group =2. For the Anxiety variable we added the Variable Label "Increase in Heart Rate", but we did not add value labels as Anxiety was a continuous variable. All of this should be familiar from Section 2 of the SPSS module of the workbook, with the exception of Variable Labels. These are a useful way to provide additional information about the nature of a variable which for convenience sake you will load under a short name. They come in particularly handy when you return to your data after a long interval, and can't remember which measure of anxiety you had used in a particular study (for just one example).

Having processed the data, we ran Explore, using Anxiety as the variable of interest and Group as the factor. This will allow us to compare numerical and graphical descriptors of Anxiety between the two experimental groups. The results of this action are presented on pages 118–119. Only the default options of Explore were used.

Examining the descriptive statistics above we note that the mean increase in heart rate for the control group in the stressful situation was greater than for the desensitization group. We also note that the latter group has a greater spread reflected in its higher variance. Both these observations are visible in both the stem-and-leaf plots and the

Explore
Therapy

Case Processing Summary

		Cases					
		Valid		Missing		Total	
	Therapy	N	Percent	N	Percent	N	Percent
Increase in Heart Rate	control	18	100.0%	0	.0%	18	100.0%
	desensitization	18	100.0%	0	.0%	18	100.0%

Descriptives

	Therapy			Statistic	Std. Error
Increase in Heart Rate	control	Mean		10.2000	.4415
		95% Confidence Interval for Mean	Lower Bound	9.2686	
			Upper Bound	11.1314	
		5% Trimmed Mean		10.2500	
		Median		10.3000	
		Variance		3.508	
		Std. Deviation		1.8730	
		Minimum		5.50	
		Maximum		14.00	
		Range		8.50	
		Interquartile Range		2.0250	
		Skewness		-.423	.536
		Kurtosis		1.672	1.038
	desensitization	Mean		5.2000	.6270
		95% Confidence Interval for Mean	Lower Bound	3.8771	
			Upper Bound	6.5229	
		5% Trimmed Mean		5.2889	
		Median		5.3500	
		Variance		7.076	
		Std. Deviation		2.6602	
		Minimum		.00	
		Maximum		8.80	
		Range		8.80	
		Interquartile Range		5.4500	
		Skewness		-.377	.536
		Kurtosis		-.966	1.038

Increase in Heart Rate

Stem-and-Leaf Plots

```
Increase in Heart Rate Stem-and-Leaf Plot for
GROUP= control

 Frequency     Stem &  Leaf

     1.00 Extremes     (=<5.5)
     3.00        8 .   249
     4.00        9 .   4678
     5.00       10 .   15688
     2.00       11 .   33
     2.00       12 .   16
      .00       13 .
     1.00       14 .   0

 Stem width:      1.00
 Each leaf:       1 case(s)


Increase in Heart Rate Stem-and-Leaf Plot for
GROUP= desensitization

 Frequency     Stem &  Leaf

     3.00        0 .   011
     3.00        0 .   223
     4.00        0 .   4555
     5.00        0 .   66777
     3.00        0 .   888

 Stem width:     10.00
 Each leaf:       1 case(s)
```

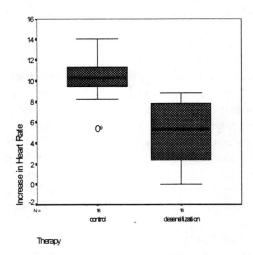

box-plots. The question is whether we have any principled reason to believe that the means would really be different if we collected another sample, or simply enlarged this one. In Chapters 5 and 6 of the text book you are presented with a method of addressing this question based on confidence intervals. You will note that SPSS Explore automatically outputs the 95% confidence intervals on the means of each group, for ease of comparison. Check Chapter 5.2 of the textbook if you are uncertain how to interpret confidence intervals. You can see that the 95% confidence intervals for the means of the two groups do not overlap, and thus we have evidence confirming our suspicion that the means really do differ. When we were activating Explore, if we had selected the Statistics button we would have been presented with a wide range of statistics which could have been output, among them a place to select the confidence interval we wish. Thus in order to obtain the 99% confidence interval on the means, we need only have changed one of Explore's options from its default setting. We now turn our attention to statistical approaches using hypothesis testing to answer this same question.

In Chapter 6 of the text we learned that an appropriate statistical method for testing the difference between the means of two groups on a given measure was the *t*-test. SPSS output uses uppercase T and so the test is referred at the "T-Test." In SPSS, the T-Test is just one way of examining such a difference, but the one which we will limit ourselves to here. As our group variable represents two independent groups of patients, we wish to use an Independent Samples T-Test. Such a test can be obtained from the **Compare Means** option, on the **Statistics** menu.

In order to produce the T-Test you have to define the grouping variable for SPSS. In the dialogue for the Independent Samples T-Test, after you have entered Anxiety as the Test variable, and Group as the Grouping Variable, you will be prompted to select the Define Groups button. This calls up another dialogue where you have several options about how to divide your data into two groups for the T-Test.

If you have a variable which is discrete, such as we have in our Group variable (remember its values are only 1 or 2) then you use the Use Specified Values option. You then tell SPSS which values are associated with which group, a value of 1 with "group 1" and a value of 2 with "group 2" in our case. Note that just because SPSS uses group 1 and 2 as the names of the levels of the grouping variable doesn't mean that your data set has to contain values of 1 and 2 for its grouping variable. It would work equally well with any other two discrete values such as characters "m" for males and "f" for females, in another data set.

The other option, "Cut Point" is used for cases where you don't have a discrete grouping variable in your data set, but a continuous variable, (or a discrete variable with more than 2 levels). Behavioral scientists are sometimes interested in comparing people who score high on some measure with those that score low. In this case our grouping variable (the test score) might be continuous, but we could treat it as though it were discrete by cutting the data at some value to divide it into two groups. It may have occurred to you that the median of a continuous variable would make an ideal candidate for such a cut point because it would break such a variable into two equal-sized groups. Such a cutting procedure is called a "median split" and breaks continuous variables into two discrete halves for use as grouping variables for T-Tests. No such elaborate procedure is needed in our case, however. Our variable of interest, Group, has only two values and thus doesn't need to be cut.

The results of selecting **Independent Samples T-Test** from the **Compare Means** option of the Statistics menu and defining Group as described above appear on page 121. Note that the output begins with a label for the variables under analysis. It is always a good idea to double check this, as we advised in Section 2. Note that once you begin conducting multiple analyses on the same or similar data sets, the helpful Variable Labels we provided in the Data menu, will become invaluable, as you struggle to keep your many pages of output organized.

T-Test

Group Statistics

	Therapy	N	Mean	Std. Deviation	Std. Error Mean
Increase in Heart Rate	control	18	10.2000	1.8730	.4415
	desensitization	18	5.2000	2.6602	.6270

Independent Samples Test

		Levene's Test for Equality of Variances		t-test for Equality of Means						
									95% Confidence Interval of the Mean	
		F	Sig.	t	df	Sig. (2-tailed)	Mean Difference	Std. Error Difference	Lower	Upper
Increase in Heart Rate	Equal variances assumed	3.228	.081	6.520	34	.000	5.0000	.7668	3.4416	6.5584
	Equal variances not assumed			6.520	30.530	.000	5.0000	.7668	3.4350	6.5650

We are next presented with some descriptive statistics for each of our groups. Double check the Ns to make sure that SPSS and you agree on the number of subjects in your groups. It is also worthwhile pausing to note which group has the larger mean, and which the larger variance at this point. Following these, there are a pair of T-Tests labeled "Equal Variances Assumed" and "Equal Variances Not Assumed" respectively. You should consult the line labeled Unequal if you have reason to believe that your two groups have different variances, and Equal otherwise. We will assume equal variances throughout. For both tests, which start in the fourth column of the chart, you are provided with a T value, and a probability of attaining a T value this extreme or greater under the null hypothesis. If the number under "sig. 2-tailed" is less than 0.05 you have reason to not retain the null hypothesis as an option in deciding whether your means are equal. In other words, If you reject the null hypothesis, you are claiming that your means are not equal, and can thus state that you have reason to believe that the group with the larger mean, truly has a larger mean than the group with the smaller mean. In this case, we are presented with a probability of 0 to 3 decimal places, which is definitely lower than 0.05 and so we conclude that the group receiving the desensitization therapy responded to the phobic situation with less increase in heart rate and thus presumably less anxiety. We have marshaled one piece of evidence in favor of belief that our therapy is successful. If you need some refreshment on the logic of hypothesis testing you should consult Section 6.4 in the textbook.

The final point of the output, contained in the second and third column of the table, presents you with a statistical test of whether your groups have equal variances. The test is called Lavene's test for equality of variances and is usually represented by F'. It Although not covered in the textbook, simply note that it is a test of the null hypothesis that the variances are equal, and thus if it is not significant at the 0.05 level we have no reason to suspect heterogeneity of variance problems, and should use the T-Test labeled "Equal Variances Assumed" above. In our example the obtained probability was given as 0.081, and so we can use the Equal Variances T value.

Review of Concepts

■ You can compare differences between two groups in your data using descriptive statistics and confidence intervals or independent samples T-tests.

■ When importing your data it is a good idea to use Variable Labels, using the Labels button in the Define Variable dialogue found in the Data menu, to give your variables longer, more descriptive names than you can use while processing them, for future reference. You should also assign Value Labels to the levels of your grouping variable, at the same time.

■ Statistics obtained from Explore can be used to compare your two groups if you previously sort the data set by the grouping variable using the factors section of the Explore dialogue. This procedure was first introduced in Section 2.

■ You can use the output of Explore to examine confidence intervals of your choosing on the means of your groups and to examine differences in the groups' means and variances numerically and graphically.

■ The Compare Means dialogue from the Statistics Menu allows you to perform an Independent Samples T-Test on the data, as well as providing useful descriptives and a test for homogeneity of variance among your groups.

Exercises

1. Import the tranquilizer data set described in Section 6.22 of your textbook. It can be found in ASCII file **DS06_01.dat**. Assign Variable Labels and Value Labels. Use Explore from the Statistics menu to examine the data from each condition, then use Compare Means to obtain a 95% confidence interval for the difference between the means of the conditions and perform an independent *t*-test. What conclusion about the effects of the tranquilizer would you draw from this study?

2. Repeat the steps described in Exercise 1 for the following data sets.

   a. The first worked example at the end of Section 6.2 (induced happiness experiment). The data are found in Data Set 6.2 (ASCII file **DS06_02.dat**).

   b. The first Problem at the end of Section 6.2 (induced anger experiment). The data are found in Data Set 6.3 (ASCII file **DS06_03.dat**).

   c. Problem 2 at the end of Section 6.2 (intentional versus incidental remembering experiment). The data are found in Data Set 6.4 (ASCII file **DS06_04.dat**).

4 ANALYSIS OF VARIANCE FOR INDEPENDENT GROUPS

Chapter 7 of the text book described the analysis of variance (abbreviated ANOVA) as a procedure for analyzing the results of experiments with more than two levels of a categorical predictor variable. In Section 7.1 you were introduced to the idea of estimating differences between the means of the levels of the predictor variable using both point estimates and confidence intervals, based on the distribution of a t statistic and on the distribution of the Studentized range statistic, q. In Section 7.2 you were introduced to a method for simultaneously evaluating differences among the group means by generating a sum of squares for the model and for the null hypothesis, dividing these sums of squares by their degrees of freedom, and taking a ratio of the resulting variances. We will begin our discussion of experiments with more than two levels using this latter technique, the analysis of variance, and then proceed to look at ways to examine differences between group means directly.

Below you will find an example of a simple ANOVA using the data presented in Section 7.1 of the text book. You can find these data in the file **DS07_01.dat**. The data reports the heart rate of phobic patients presented with their phobic object, after receiving one of three approaches to treatment. For the purposes of this example we will call the predictor (independent) variable Group and the response variable Heartrt. To perform a one-way ANOVA you select the **Compare Means** item from the **Statistics** menu. From the available options select **One-Way ANOVA....** This will bring up a dialogue with the same label, which will allow you to set up the model you wish to test. From your list of variables, presented on the left hand side of the dialogue select **heartrt** (or whatever you've decided to call it) and move it to the box labeled **Dependent List**, by pressing the arrow next to this box. Then select **condit** (or whatever you've named the class variable) and move it to the box labeled **Factors** in the same way. You now have several options available. You can run the analysis immediately using the **OK** button, or you can elect to have some additional statistics calculated at this time by using one of the buttons at the bottom of the dialogue.

Pressing the **Options**... button brings up a dialogue with two parts. The bottom half offers two methods for SPSS to handle missing values. It is usually advisable to leave this set to the default setting. The top half of this dialogue includes two check boxes, one requesting **Descriptives** and the other a test of **Homogeneity-of-variance**. For the following output we have selected both. Press the Continue button to return to the One Way ANOVA dialogue.

Oneway

Descriptives

			N	Mean	Std. Deviation	Std. Error	95% Confidence Interval for Mean		Minimum	Maximum
							Lower Bound	Upper Bound		
H_RATE	CONDIT	1.0000	15	10.0000	2.478479	.639940	8.627464	11.3725	4.0000	14.0000
		2.0000	15	6.000000	2.507133	.647339	4.611596	7.388404	2.0000	10.0000
		3.0000	15	5.000000	2.618615	.676123	3.549860	6.450140	1.0000	9.0000
		Total	45	7.000000	3.302891	.492366	6.007702	7.992298	1.0000	14.0000

Test of Homogeneity of Variances

	Levene Statistic	df1	df2	Sig.
H_RATE	.380	2	42	.686

ANOVA

		Sum of Squares	df	Mean Square	F	Sig.
H_RATE	Between Groups	210.000	2	105.000	16.333	.000
	Within Groups	270.000	42	6.429		
	Total	480.000	44			

The first table produced lists the descriptive statistics we requested. It is useful to have a table of means, standard deviations, etc. close to hand when you interpret the results of your ANOVA. Thus it is virtually always a good idea to select this option, whenever you run an analysis of variance. It should go without saying, however, that you should have examined your variables before even beginning this analysis, using the various techniques discussed in Section 2 of this workbook.

Second in our output, we have a printout of the test for homogeneity of variance. This is a formal test, using an *F* statistic itself, which compares the variances of the three groups in our data. The test is run against a null-hypothesis that the variances of our three groups are equal (much as a normal ANOVA tests the null-hypothesis that the means of our groups are equal). The fact that this test has turned up non-significant is thus a good thing. It means that we have met the homogeneity of variance assumption underlying any analysis of variance.

Finally, we have the table summarizing the results of our ANOVA itself. The form of the table should be familiar from Chapter 7 of the text. The source labeled as Between Groups reports the sum of squares for our model, while the source labeled Within Groups reports the sum of squares which is residual variance. After reporting the sums of squares, degrees of freedom, mean squares and the *F*-ratio itself, the program reports the exact probability associated with an *F* of this magnitude. In our example, it has listed a probability of .000 which does not mean 0 but rather a very small number, beyond the significant digits which are printed by default in the output from the ANOVA. To be conservative in reporting your results you could report this minuscule p-value as <0.001. The correct format for reporting this statistical test in the psychological literature would be to state that we have obtained: $F(2,42) = 16.333$, $p<0.001$.

The significant results obtained for Group, in the above output only tells us that the means of the three groups are not the same. It tells us nothing about which means are larger, or indeed which means differ from one another. To further examine questions of differences among the treatment means, it is necessary to engage in a comparison procedure which will control the error rate per experiment. The textbook identified two approaches to this problem. One could test the differences among the means independently of any ANOVA using either a Bonferroni procedure to control the Type-1 error rate for each t-test, or Tukey's use of the Studentized range statistic, q, to perform simultaneous comparisons. The second option, was to run the ANOVA, and then if one obtained significant results, to examine comparisons among pairs of means, using t-tests. This second procedure was known as Fisher's LSD. Unfortunately this distinction is somewhat blurred in SPSS by the fact that any of these comparisons can only be requested during the request that an ANOVA be run. To request a comparison of group means one merely selects the **Post-Hoc...** button at the bottom of the One-Way ANOVA dialogue. This brings up a new dialogue which lists numerous procedures for comparing group means.

In this case, we have selected Tukey (the SPSS name for Tukey's HSD), Bonferroni and LSD to allow for comparison between these three procedures. Normally you would only select one, however. In addition to the two tests we have discussed, we have included a command for a third comparison procedure called REGWQ in SPSS for the first initials of the four statisticians who have contributed to its development: Ryan, Einot, Gabriel and Welch. From here on we will refer to it as the Ryan procedure, after the person who first suggested this approach. This method is not covered in the textbook, but is mentioned here for those who may be interested. The Ryan procedure is similar to Tukey's HSD in that it tests means based on a value of the Studentized range statistic, q. Unlike Tukey, however, it uses a different range for each comparison and in this way is like the older, now discredited, Newman-Keuls procedure (which we could also have selected). Unlike that earlier procedure, however, Ryan adjusts the alpha level of each comparison to take into account the number of comparisons being made. It thus offers an excellent compromise between the Tukey procedure (which most feel is too conservative) and Fisher's LSD or Newman-Keuls (which most feel are too lenient). It is likely to become more popular in coming years, though it is too computationally cumbersome to carry out by hand. As of this writing SAS and SPSS are the only statistical packages to support this innovative technique, and thus we present it for your consideration, along with the more traditional techniques.

Fisher's LSD is reported as controlling the error rate per comparison, rather than per experiment. This indicates that with large numbers of comparisons (ie. A study with many levels of the grouping variable), the alpha rate for the experiment may become unacceptably high. Tukey's HSD controls the error rate per experiment by furnishing a minimum difference between means based on the largest range of ordered means possible in the experiment. While this controls Type-1 errors, it is too conservative and thus we should be warned of the increased chance of making a Type-2 error. The Ryan procedure controls error rate per experiment by furnishing a minimum difference between means based on the range of each pair of ordered means, using a Bonferroni-like procedure to adjust the alpha level for each comparison. Thus control over the possibility of Type-1 errors is maintained while not increasing the likelihood of a Type-2 error. The results of this selection appear on page 126.

Post Hoc Tests

Multiple Comparisons

Dependent Variable: H_RATE

	(I) CONDIT	(J) CONDIT	Mean Difference (I-J)	Std. Error	Sig.	95% Confidence Interval	
						Lower Bound	Upper Bound
Tukey HSD	1.0000	2.0000	4.000000*	.926	.000	1.750719	6.249281
		3.0000	5.000000*	.926	.000	2.750719	7.249281
	2.0000	1.0000	-4.000000*	.926	.000	-6.24928	-1.75072
		3.0000	1.000000	.926	.532	-1.24928	3.249281
	3.0000	1.0000	-5.000000*	.926	.000	-7.24928	-2.75072
		2.0000	-1.000000	.926	.532	-3.24928	1.249281
LSD	1.0000	2.0000	4.000000*	.926	.000	2.131619	5.868381
		3.0000	5.000000*	.926	.000	3.131619	6.868381
	2.0000	1.0000	-4.000000*	.926	.000	-5.86838	-2.13162
		3.0000	1.000000	.926	.286	-.868381	2.868381
	3.0000	1.0000	-5.000000*	.926	.000	-6.86838	-3.13162
		2.0000	-1.000000	.926	.286	-2.86838	.868381
Bonferroni	1.0000	2.0000	4.000000*	.926	.000	1.691313	6.308687
		3.0000	5.000000*	.926	.000	2.691313	7.308687
	2.0000	1.0000	-4.000000*	.926	.000	-6.30869	-1.69131
		3.0000	1.000000	.926	.859	-1.30869	3.308687
	3.0000	1.0000	-5.000000*	.926	.000	-7.30869	-2.69131
		2.0000	-1.000000	.926	.859	-3.30869	1.308687

*. The mean difference is significant at the .05 level.

Homogeneous Subsets

H_RATE

	CONDIT	N	Subset for alpha = .05
			1
Tukey HSD[a]	3.0000	15	5.000000
	2.0000	15	6.000000
	1.0000	15	
	Sig.		.532
Ryan-Einot-Gabriel-We lsch Range	3.0000	15	5.000000
	2.0000	15	6.000000
	1.0000	15	
	Sig.		.286

Means for groups in homogeneous subsets are displayed.

a. Uses Harmonic Mean Sample Size = 15.000

The output breaks down our requested statistics into two parts. The first reports the results of the Tukey's HSD, the Bonferroni-adjusted t-tests and Fisher's LSD. In this section, point estimates of the differences between the group means are reported and tested for significance, with significant differences marked with asterisks. Additionally, the program reports 95% confidence intervals on these point estimates based upon the Studentized range statistic, q. in the first instance, and the t-statistic in the next two. The use of this information should be apparent from the discussion of such confidence intervals in Section 7.1 of the textbook.

The second part of the output reports the results for the comparison procedures using the Studentized range statistic, q, in a different format. Here we see the results of Tukey's HSD and the Ryan procedure reported as a list of the levels of group in descending order of magnitude of the group means. In the rightmost column are reported the group means of any adjacent groups which do not differ from one another. A quick inspection of these results indicates that all the methods we selected are in agreement that the mean of group 3 is significantly different from the means of groups 1 and 2, and that the means of these latter groups do not differ from each other.

Factorial ANOVA

In Section 7.3 of the textbook you were introduced to a method for testing differences among means in experiments involving two (or more) predictor variables using Factorial ANOVA. In a two way ANOVA we assume that each data point can be modeled by a linear combination of a main effect for each of the two predictor variables plus an interaction term for the effects of the unique combinations of these predictor variables, plus , of course, a residual component. An example of such an experiment is given in Section 7.3, involving the effect on driving performance of two different levels of sleep deprivation, in combination with two different dosages of alcohol. The data for this experiment can be found in the data file **DS07_07.dat**.

To test this data we select **Simple Factorial** from the **General Linear Model** menu item on the **Statistics** menu [Note that in the student edition you will find **Simple Factorial** under the **Anova Models** menu item]. This will call up a dialogue, in which you are presented with a box containing a list of your variables and three other boxes, labeled **Dependent**, **Factor(s)**, and **Covariate(s)** respectively. Using the arrow buttons, move the variable for reaction time (we called it Time in our data set) to the **Dependent** box, and both of the predictor variables (we labeled then Alcohol and Sleep in our example) to the **Factor(s)** box. The third box is used in a specialized form of Factorial design called Analysis of Covariance or ANCOVA which is beyond the scope of our text. After you have moved your variables you will note that there are question marks in brackets behind the variable names you moved to the **Factor(s)** box. SPSS is waiting for you to define the ranges of these factors (i.e., how many levels there are of each variable). By selecting each of your response variables in turn and then pressing the **Define Ranges...** button, you will be presented with a dialogue in which you can enter the minimum and maximum values of each of your predictor (indepedent) variables (1 and 2 respectively in this case). Once you have done this you can press the **OK** button to continue with the analysis, leaving all the default options in place. The results are shown on page 128.

As can be seen both of our main effects are significant, but because we are dealing with a factorial design, we must first check to see if our interaction is significant. It is, and therefore we will have to interpret our main effects in light of this interaction. One way to do this is to test for simple effects. Simple effects are the differences between the means of one of our predictor variables, tested at each level of the other predictor variable—for example, the differences between the two levels of sleep deprivation for each

ANOVA

Case Processing Summary[a]

	Cases					
	Included		Excluded		Total	
N	Percent	N	Percent	N	Percent	
40	100.0%	0	.0%	40	100.0%	

a. TIME by ALCOHOL, SLEEP

ANOVA[a,b]

			Unique Method				
			Sum of Squares	df	Mean Square	F	Sig.
TIME	Main Effects	(Combined)	1.258	2	.629	25.104	.000
		ALCOHOL	.729	1	.729	29.095	.000
		SLEEP	.529	1	.529	21.113	.000
	2-Way Interactions	ALCOHOL * SLEEP	.196	1	.196	7.823	.008
	Model		1.454	3	.485	19.344	.000
	Residual		.902	36	2.5E-02		
	Total		2.356	39	6.0E-02		

a. TIME by ALCOHOL, SLEEP

b. All effects entered simultaneously

level of alcohol consumption. In other words we wish to perform two more one-way ANOVAs on this data, using half the data for each.

To divide our data into two groups based on Alcohol (for example) we use the **Split Files...** menu option from the **Data** menu. This brings up a dialogue in which we tell SPSS how to split the cases of our original data into two (or more) groups. Select **Organize output by groups** [this option is called **Repeat analysis for each group** in the student edition] and then use the arrow button to move the variable of your choice (in our case Alcohol) to the window labeled **Groups Based on**. In order to split a file by a variable it must first be sorted by that variable. This is not a separate operation you have to perform however, as just leaving the default option of **Sort the file by groups variable** will cause this to happen. Note the other option is there to save time and processing power if you know that the file is already sorted, but in most circumstances it is best to leave the defaults alone. Better safe than sorry. Once you have finished with the **Split Files** dialogue, there will be a note saying **Split File On** at the lower right hand corner of your screen. This is to remind you that anything you do (statistics, graphs, etc.) from this point on will be performed separately on the data from each subset of your grouping variable (Alcohol). Whenever you wish to use the whole dataset again, you just return to the **Split File** dialogue, and turn the grouping off by selecting **Analyze all cases, do not create groups**.

After we have split our file by the two levels of Alcohol, we are ready to conduct our one-way ANOVAs. Simply follow the procedures outlined above, selecting Sleep as your factor and Time as your response variable, and you will automatically receive the two outputs, one for each level of Alcohol. Note that we have also selected **Descriptives** from the **Options** button. The results follow:

Oneway
ALCOHOL = 1.0000

Descriptives[a]

			N	Mean	Std. Deviation	Std. Error	95% Confidence Interval for Mean		Minimum	Maximum
							Lower Bound	Upper Bound		
TIME	SLEEP	1.0000	10	.680000	.193218	6.1E-02	.541780	.818220	.4000	1.0000
		2.0000	10	.770000	.163639	5.2E-02	.652940	.887060	.5000	1.1000
		Total	20	.725000	.180278	4.0E-02	.640628	.809372	.4000	1.1000

a. ALCOHOL = 1.0000

ANOVA[a]

		Sum of Squares	df	Mean Square	F	Sig.
TIME	Between Groups	4.0E-02	1	4.0E-02	1.263	.276
	Within Groups	.577	18	3.2E-02		
	Total	.617	19			

a. ALCOHOL = 1.0000

ALCOHOL = 2.0000

Descriptives[a]

			N	Mean	Std. Deviation	Std. Error	95% Confidence Interval for Mean		Minimum	Maximum
							Lower Bound	Upper Bound		
TIME	SLEEP	1.0000	10	.810000	.128668	4.1E-02	.717956	.902044	.6000	1.0000
		2.0000	10	1.180000	.139841	4.4E-02	1.079964	1.280036	.9000	1.4000
		Total	20	.995000	.230503	5.2E-02	.887121	1.102879	.6000	1.4000

a. ALCOHOL = 2.0000

ANOVA[a]

		Sum of Squares	df	Mean Square	F	Sig.
TIME	Between Groups	.685	1	.685	37.911	.000
	Within Groups	.325	18	1.8E-02		
	Total	1.010	19			

a. ALCOHOL = 2.0000

As can be seen we have two complete outputs from our ANOVA procedure, one for the scores at each level of Alcohol consumption. By examining these simple effects analyses, the real pattern of results, giving rise to our initial interaction term is apparent. There is an effect of sleep deprivation (at the two levels employed) but only when subjects have consumed alcohol.

Review of Concepts

■ The procedure for testing the differences among the means of more than two groups simultaneously is the analysis of variance (ANOVA).

■ To perform a simple, or one-way ANOVA in SPSS you select the **Compare Means** item from the **Statistics** menu. From the available options you select **One-Way ANOVA....**

■ In addition to the overall ANOVA you can request a printout of descriptive statistics and a test of homogeneity of variance from the **Options...** button on the One-Way ANOVA dialogue. Both of these options are highly recommended.

■ To explore differences among the means directly you can select a variety of post-hoc comparisons by selecting the **Post-Hoc...** button on the one-way ANOVA dialogue.

■ Of the various options presented, several should be familiar to readers of the text, and in additon we recommend a relatively new procedure called REGWQ as a good compromise between control of the error rate per experiment and power for each comparison.

■ The post hoc procedures report their results in the form of point estimates with attached significance tests, and in the form of 95% confidence intervals on those estimates.

■ ANOVA models involving more than one predictor variable can be tested using the **Simple Factorial** item from the **General Linear Model** item on the **Statistics** menu.

■ Significant interactions can be most easily interpreted using simple effects analysis. This involves performing separate one way ANOVA tests on one of your predictor variables using the part of your data at each level of the other predictor variable. You can subdivide your data in groups for this purpose (or turn grouping off, later) using the **Split Files...** item from the **Data menu**.

Exercises

1. Use the data from **DS07_02.dat**. These data examining tranquilizer effects were used in the first worked example in Section 7.1 of the Text. Compare the means the three conditions using an one-way ANOVA. Further examine differences among the Groups using Tukey's HSD. Finally, have SPSS produce 95% confidence intervals for the Group differences, based upon the t statistic (Fisher's LSD).

2. Repeat the steps in Exercise 1 to the following data sets.

   a. **DS07_03.dat** (see Worked Example 2 in Section 7.1 of the text).

   b. **DS07_04.dat** (see Problem 1 in Section 7.1 of the text).

   c. **DS07_10.dat** (see Worked Example 2 in Section 7.3 of the text). This experiment requires a factorial ANOVA. Obtain a complete analysis of variance summary table and test the significance of main effects and the interaction.

5 MATCHED PAIRS AND WITHIN-SUBJECTS DESIGN

Matched Pairs for Two Conditions

In Chapter 8 of the text a method was introduced for controlling extraneous sources of variance by matching subjects (or testing the same subjects) on the two conditions of an experiment. The net effect of this procedure was to reduce the variance attributable to residuals, when testing our model, by making the variance attributable to individual differences among subjects an explicit part of our model. To perform this operation for an experiment with two conditions, we first calculated difference scores for each subject, which were simply the subjects' score on the first condition minus the subjects' score on the second condition. We then tested the difference between the conditions by testing the null hypothesis that the mean of these difference scores was in fact 0.

To perform this procedure in SPSS we will not need to add a difference variable to our data. SPSS has built in procedures for handling pairs of variables, and testing the differences between them. We will use the data presented in Section 8.2 of the text book, which presents a matched samples version of the experiment introduced in Chapter 6. The data for this experiment can be found in the file DS08_01.dat.

To perform our analysis we select **Compare Means** from the **Statistics menu**. We then select **Paired Samples T Test...** from the available options. This will call up a dialogue in which you can tell SPSS which variables to pair. Select the variable for your first condition from the list of available variables at the top left of the dialogue. That name will immediately appear next to **var 1** at the bottom left. You now repeat the procedure for the variable listing the scores for the second condition. These will appear next to **var 2**. If this is the pair you wish to test, press the arrow button to move the pair to the **Paired Variables** list box in the middle of the dialogue. Pressing the **OK** button will produce the output shown on page 132.

The first table lists descriptive statistics for the two conditions of our experiment. The second table presents a Pearson's product moment correlation test for the two conditions. Consult Chapter 9 of the text if you are uncertain of its interpretation. For now, note simply that a significant result on this test certainly provides justification for our having treated these data as a paired samples t-test rather than an independent groups t-test. The third table presents the results of our actual t-test. As can be seen, the value for the t score of approximately 2.41 agrees with the value calculated in the textbook, and the probability value of 0.026 is certainly less than our usual designated alpha level of 0.05, thus we have reason to reject the null hypothesis. Remembering that the null hypothesis we are testing is that the mean of the differences between the conditions is 0,

T-Test

Paired Samples Statistics

		Mean	N	Std. Deviation	Std. Error Mean
Pair 1	EXPMNT	71.0000	20	7.940569	1.775565
	CONTROL	67.0000	20	7.953814	1.778527

Paired Samples Correlations

		N	Correlation	Sig.
Pair 1	EXPMNT & CONTROL	20	.563	.010

Paired Samples Test

		Paired Differences							
					95% Confidence Interval of the Difference				
		Mean	Std. Deviation	Std. Error Mean	Lower	Upper	t	df	Sig. (2-tailed)
Pair 1	EXPMNT - CONTROL	4.000000	7.426836	1.660691	.524134	7.475866	2.409	19	.026

we can thus conclude that we have reason to believe there is a real difference between the conditions of our experiment. We will now examine the question of experiments with more than 2 conditions. As you might expect, we will have to shift our test statistic of choice from t to F, as we more to more than 2 conditions.

Repeated Measures for More Than Two Conditions

The method for testing the null hypothesis that the means of more than two groups are the same, when we are sampling the same (or closely matched) subjects in each of our conditions may at first seem somewhat confusing. It is helpful to remember what we are attempting to do. In choosing to use a repeated measures methodology, we are attempting to reduce the unaccounted for variance (residual mean square) by making the variance due to individual differences between our subjects explicit in our model. Accordingly, even though we are only interested in one variable, our grouping variable, we will have to construct an ANOVA model as though we were interested in two, our grouping variable and the differences among the subjects. We accomplish this by using a variable containing an identifier of the subjects as part of our model, denoted in the text by π_j.

For our example we will be using the data from Section 8.3 of the text book. In this study, 17 subjects were tested in each of three conditions of testing time preceding and following the attribution manipulation. These data can be found in the file: **DS08_07.dat**. It consists of three columns of data, which we have labeled subject, attrib and score, respectively.

The ANOVA procedure should look fairly similar to examples from the last chapter, with two notable exceptions. As we mentioned in order to eliminate the variability among subjects from our residual term, we have to make it explicit in our model. Thus we list both Attrib and Subject in our model.

In order to conduct this analysis first select the **Statistics** menu, and from it select the **General Linear Model** item. From the list of alternatives, select **GLM: General Factorial...** if you have that option available. [Note, if you are using the student edition of SPSS you will have to make do with using the **Simple Factorial** option from the **Anova Models** menu item, instead. The only difference is that you won't be able to designate subject as a random factor, and you won't have the same options for additional output, like post-hoc tests. Otherwise follow the same methodology here.]

This brings up a dialogue from which you can specify all the factors in our ANOVA model, plus request additional tests of differences among the means. The layout of this dialogue should be quite familiar by now. To the left is a list of the variables in our data file, and to the right are a number of boxes you can move the variables to, using the adjacent arrow buttons. Move score to the **Dependent Variable** box, attrib to the **Fixed Factor(s)** box, and subject to the **Random Factor(s)** box. We call subject a random factor because we have randomly selected the subjects for our study from a pool of potential subjects. If we were to repeat this experiment it would be extremely unlikely that we would use any of the same subjects. [Note that if you are using the student version, there is only one factors box, and you will have to put both attrib and subject here.]

The next step is the most important. Press the **Model...** button. This will bring up a dialogue for you to further specify the model that you are testing. At the top are a pair of buttons labeled **Fully Factorial** and **Custom**. Pick the custom button. You now can access the boxes below, with the adjacent arrow button exactly as before. Move both attrib and subject(R) into the **Model** box. This will create an ANOVA with a test of a treatment called attrib and a treatment called subject. Recalling factorial designs, you might well ask why we have not included a third term in the model, representing the interaction of Attrib and Subject. (an interaction represents the unique contribution to the variability of our scores caused by specific combinations of our two predictor variables.) The answer to this question is simple, with only 1 subject's score present in the cell at each combination of Attrib and Subject, there is no way to separate the variability due to the interaction from the residual variability within cells. The two sources of variation are completely confounded. Our residual term, *is* the interaction term. This is why we didn't let the program create a fully factorial model, as it would have tried to fit an interaction treatment called attrib x subject, and there would have been no variability left for an residual term. [Note, if you are using the student version you can suppress the creation of this interaction term using the **Options...** button. On the subsequent dialogue, find the place in the lower right where it asks you for the maximum level of interaction it should test, and pick no interaction, as your option. You will thus be performing a similar test to those using the full version of SPSS]. Also while on this dialogue you probably want to suppress the testing of an intercept in the model. All the intercept term tests is the null hypothesis that the grand mean is 0. This is not usually something we care about, and thus testing the intercept simply increases your error rate per experiment for little information gained. Simply remove the tick mark from the box at the bottom of the dialogue.

Having created our model we can now select other tests we would like SPSS to carry out. From the dialogue raised by pressing the **Post Hoc...** button you can select any post hoc comparisons you wish to have carried out on the levels of attrib. For our example we have selected the Ryan procedure described in the last section. You can also choose to have SPSS display marginal means in a table using the **Options...** button (often useful for the interpretation of the results of your ANOVA) or in a plot using the **Plots...** button. Also on the options dialogue you can request a display of effect size for each of your tests (which also includes an analysis of power).

Pressing **OK** will begin your analysis. After listing the factors, the results from these selections will appear as follows.

Tests of Between-Subjects Effects

Dependent Variable: SCORE

Source		Type III Sum of Squares	df	Mean Square	F	Sig.	Eta Squared	Noncent. Parameter	Observed Power[a]
ATTRIB	Hypothesis	127.216	2	63.608	7.048	.003	.306	14.097	.904
	Error	288.784	32	9.025[b]					
SUBJECT	Hypothesis	358.745	16	22.422	2.485	.014	.554	39.752	.943
	Error	288.784	32	9.025[b]					

a. Computed using alpha = .05

b. MS(Error)

Expected Mean Squares[a,b]

	Variance Component		
Source	Var(SUBJECT)	Var(Error)	Quadratic Term
ATTRIB	.000	1.000	ATTRIB
SUBJECT	3.000	1.000	
Error	.000	1.000	

a. For each source, the expected mean square equals the sum of the coefficients in the cells times the variance components, plus a quadratic term involving effects in the Quadratic Term cell.

b. Expected Mean Squares are based on the Type III Sums of Squares.

Estimated Marginal Means

ATTRIB

Dependent Variable: SCORE

ATTRIB	Mean	Std. Error
1.0000	15.2941	.729
2.0000	18.2353	.729
3.0000	18.9412	.729

Most of the preceding output should be fairly self-explanatory. The first table can be safely ignored, it is simply SPSS' way of confirming the levels of your predictor variable(s), you can double check this if your results seem unusual. Next we have our ANOVA summary table. Don't be confused by the fact that SPSS refers to both subjects and attrib as "between subjects effects" this is just a residue of the method of testing we have had to employ (See the end of this section for a fuller discussion). Simply note that the mean squares, and tests conducted here will be accurate for our "within subjects variable", attrib. Note that we also have a significant effect reported for subject. As with the correlation coefficient reported during the paired samples t-test, this is not of primary interest to us, but rather serves as an indication that our decision to run the experiment as a repeated measure was warranted. Subjects do indeed differ from one another on our response measure, and thus if we had not run this as a repeated measures ANOVA, all that extra variability would have ended up in the residual term of our one-way ANOVA.

To examine the differences among our three means, we can now turn to the results of our post-hoc test(s). Note that although we have chosen to use the Ryan procedure again, you should feel free to use Tukey's HSD or Fisher's LSD or whichever test you prefer. Simply refer to the previous section of the workbook for a reminder on how to invoke any of these other tests. The results for this test show that the means of Attrib condition 3 (Delayed Post test) and condition 2 (Immediate Post test) are both significantly greater than the mean of condition 1 (Pretest) but that they do not differ from each other. In other words the attribution manipulation worked, with evidence of learning being apparent immediately after the manipulation, and still apparent some time afterwards.

Post Hoc Tests
ATTRIB
Homogeneous Subsets

SCORE

Ryan-Einot-Gabriel-Welsch Range[a]

ATTRIB	N	Subset 1
1.0000	17	
2.0000	17	18.2353
3.0000	17	18.9412
Sig.		.498

Means for groups in homogeneous subsets are displayed.
Based on Type III Sum of Squares

a. Alpha = .05.

Review of Concepts

■ Using matched samples or the same subjects are methods of reducing the variance in our models that we have to residuals, by making the individual differences between subjects explicit.

■ When you have an experiment with two conditions the test of choice is a t-test performed on difference scores calculated between our conditions.

■ To perform such a test in SPSS you need not first create the new dependent variable in your data set. You simply select **Compare <u>M</u>eans** from the **<u>S</u>tatistics** menu and then select **<u>P</u>aired Samples T Test...** from the available options.

■ In the event that we have more than two conditions in an experiment we will rely on a repeated measures ANOVA as our test.

■ Performing such a test in SPSS is virtually identical to the one way ANOVA we discussed in the last section. You just need a variable coding the identity of subjects in you data set and then you add this term for subjects to your model. Depending on whether you are using a full or student version of SPSS you can use either the General Factorial or the Simple Factorial options. The important thing is to create a model with both you subject term and your independent grouping variable of interest, but without an interaction term. You then examine the results of your response variable exactly as you would have in Section 4.

Exercises

From Chapter 8 of the textbook, try to use SPSS to answer the questions in Exercise 5 and Problems 1 and 2. In addition to the questions asked, perform a post-hoc test (of your choice) on the differences between the group means in each problem.

6 REGRESSION AND CORRELATION

Linear Regression

In Chapter 9 of the text book you were introduced to a statistical technique called linear regression. Unlike the techniques described in the previous chapters, regression has the advantage of allowing you to develop a model describing a functional relationship between two (or more) variables. In the textbook this equation was written as $\hat{Y} = \alpha + \beta X$, where α is the intercept and β is the slope. The methodology you learned in Chapter 9 obtains estimates (a and b) of the parameters a and b. On the basis of these values you construct an equation from which you can predict the values of Y by multiplying the estimated slope by the values of X and then adding estimated value of the intercept. We can test how well our line fits the data, first by calculating sums of squares for our regression line, and then for the residuals (differences between our predicted values of Y and our obtained values of Y). The result is an ANOVA table testing the null hypothesis that $\beta = 0$.

We can further examine our residuals by plotting them against values of our response measure and against our predicted values to look for any systematic deviations from the straight line our model predicts. As an example of this methodology we will be using the first data set found in Section 9.1 of the textbook. These data can be found in file **DS09_01.dat**. Our response variable is the number of errors made by subjects under varying conditions of sleep deprivation, while our predictor variable is the number of hours of sleep deprivation experienced by that subject. In other words our Y variable is Errors, while our X variable is Hours. We will now construct a model specifying a linear relationship between these variables and then test that model using an ANOVA, and using graphical analyses of the residuals from our model.

In order to perform this analysis in SPSS you first select the **Statistics** menu, and then select the **Regression** item. From the many alternative regression models available select **Linear...** (the other models are useful in situations not covered in this book). Making these selections calls up a dialogue in which you can specify the model you wish to test, as well as any additional tests you wish SPSS to carry out. The selection boxes in the middle of the dialogue should be familiar to you by now. Select Errors from your list of variables and move it to the **Dependent Variable** box, using the adjacent arrow button. Similarly move Hours to the **Independent Variable(s)** box.

The **Statistics...** button takes you to a dialogue where you can select the output you would like from the regression analysis. The default value of **Estimates** will produce a table of our two regression coefficients (the slope and the intercept) while the default analysis of **Fit** will produce our ANOVA summary table testing the null hypothesis that all our coefficients are zero.

The **Plots...** button takes you to a dialogue from which you can request your residual plots, to graphically aid in your analysis of your model's adequacy. To select a plot you simply choose one of the variables to the left (note that most of these are new variables computed as part of SPSS' regression program) to move to the **X axis** box and another to move to the **Y axis** box. If you wish to request a second plot (or a third, fourth, etc.) you simply press the **Next** button, the two boxes will be blanked, and the label will toggle from **Scatter 1 of 1** to **Scatter 2 of 2** (or whatever). You can always change a plot you specified earlier by pressing the **Previous** button until the desired variables reappear in the boxes. In general you only want a plot of the predicted values (on the Y axis) versus your original response variable (on the X axis), and a plot of your residuals (on the Y axis) versus your predicted values (on the X). Using SPredicted and SResidual will use the respective variables transformed by the *t*-transformation (and thus Studentized). More easily readable are the values of your variables transformed by the *z*-transformation, for which you would select ZPredicted and ZResidual, respectively. In order to obtain just the two plots we have suggested and using *z*-transformed predicted values and residuals, you can simply check the box marked **Produce all Partial Plots**. The results of conducting the above analysis using the settings suggested appear on page 139.

The first item produced is a rather ominously labeled warning box. Careful reading, however, will reveal that we have no problems with our data, and thus in this case the box only serves as a place holder for real warning messages. If we had problems with our data, for example some subjects missing a value for hours, or with outliers for hours or errors, we would have been told of this here, and informed that those cases should be dropped from the analysis. Outliers can present a serious problem in regression analysis as they can strongly influence the slope of the regression line. Imagine the effect of just one point that had a value of hours 10 times the next highest value. It would exert undue influence on the slope of the line simply because of its position relative to the majority of the data. Such points are called high leverage points. (Try changing one of the data points in this way and see what happens).

Next we have a table labeled model summary. It is here that we get our value of R-square (R^2) for our model, along with a standard error on this estimated value, enabling us to place confidence intervals around the estimate. R-square is our measure of association (or alternately of effect size). Our value of 0.594 means that we can account for approximately 60% of the variance in subjects error score based upon our knowledge of the number of hours they have been deprived of sleep. It thus presents an index of the predictive power of our model. Note that there is an additional value called Adjusted R-square. In general as you use more predictor variables in your model, the value of R-square will become positively biased, purely as an artifact of the way we calculate our model sums of squares. Accordingly it becomes increasingly desirable to have an estimate of degree of association that is unbiased as we add more predictors. Note that with only the one predictor variable there is virtually no difference between the two values.

Next we have the ANOVA summary table for our model. Note that the sums of squares, degrees of freedom, etc. correspond to those in the textbook. Our regression has an *F*-value > 70, which in turn we are told is significant at a *p*-value <0.001. We can thus safely discredit the null hypothesis that our coefficients = 0 and thus have one more reason to believe that our model fits the data well.

Finally we have the table listing our model coefficients. According to this table our best model for predicting errors is: errors = 0.671 (hours) + 13.53. In addition to the point estimates of our slope and intercept we are given a standard error on each esti-

Regression

Warnings

For the final model with dependent
variable ERRORS, no outliers were
found. No casewise diagnostics
are produced.

Model Summary[a,b]

Model	Variables		R	R Square	Adjusted R Square	Std. Error of the Estimate
	Entered	Removed				
1	HOURS[c,d]	.	.771	.594	.586	6.406101

a. Dependent Variable: ERRORS

b. Method: Enter

c. Independent Variables: (Constant), HOURS

d. All requested variables entered.

ANOVA[a]

Model		Sum of Squares	df	Mean Square	F	Sig.
1	Regression	2883.690	1	2883.690	70.269	.000[b]
	Residual	1969.830	48	41.038		
	Total	4853.520	49			

a. Dependent Variable: ERRORS

b. Independent Variables: (Constant), HOURS

Coefficients[a]

Model		Unstandardized Coefficients		Standardized Coefficients	t	Sig.	95% Confidence Interval for B	
		B	Std. Error	Beta			Lower Bound	Upper Bound
1	(Constant)	13.530	2.125		6.368	.000	9.258	17.802
	HOURS	.671	.080	.771	8.383	.000	.510	.832

a. Dependent Variable: ERRORS

mate and a 95% confidence interval calculated around each of these estimates. Additionally there is a column for standardized coefficients. These are the coefficients which would have been produced had all our variables been standardized (z-transformed) before we ran our regression. These values are useful when you have more than one predictor as they allow you to directly compare the sizes of your various slope coefficients. Note that there is no standardized estimate of the intercept. If all the scores had been standardized our regression line would have passed through the origin (giving us an intercept of 0).

Following these three tables are two further outputs. The first lists a fuller description of any problems identified in the initial warning message. As we had no problematic data there are no Casewise Diagnostics listed. The second is a chart giving descriptive statistics on our model residuals. It lists the minimum, maximum and mean predicted values along with the residuals associated with each, for both natural and standardized scores. The two tables follow.

Casewise Diagnostics[a]

a. No outliers were found for one or more split files.

Residuals Statistics[a]

	Minimum	Maximum	Mean	Std. Deviation	N
Predicted Value	18.90000	40.38000	29.6400	7.671429	50
Residual	-14.2700	12.36000	-1.3E-15	6.340395	50
Std. Predicted Value	-1.400	1.400	.000	1.000	50
Std. Residual	-2.228	1.929	.000	.990	50

a. Dependent Variable: ERRORS

Finally, we can examine our two plots for any problems with our model. You will recall that the first chart we requested was a plot of our standardized predicted values against our response measure. Second we requested a plot of our standardized residuals against our standardized predicted values. The charts we selected can be seen below. In the first chart we are looking for any deviation from a linear relationship. If there were any curves apparent in this chart it would be evidence that the best model for our data was probably not a linear model. At that point we would have several options, the most commonly exercised being a return to the data set, transforming either or both variables in an attempt to linearize their relationship and then a new regression analysis on the transformed data. The second chart serves a somewhat similar function. It is a test of the adequacy of our model across its range. The residuals should be dispersed fairly evenly to either side of the zero line, along the length of the x axis. If there were any patterns evident in their dispersal we would again have evidence of problems with our model and would probably have to adopt a strategy similar to that for problems with the first chart (see page 141).

As can be seen, there are no serious problems with either chart, and thus we can probably safely consider our model a good one. We now have three pieces of evidence, the value of R-square, the significant ANOVA and the residual plots indicating a good model has been selected.

Charts

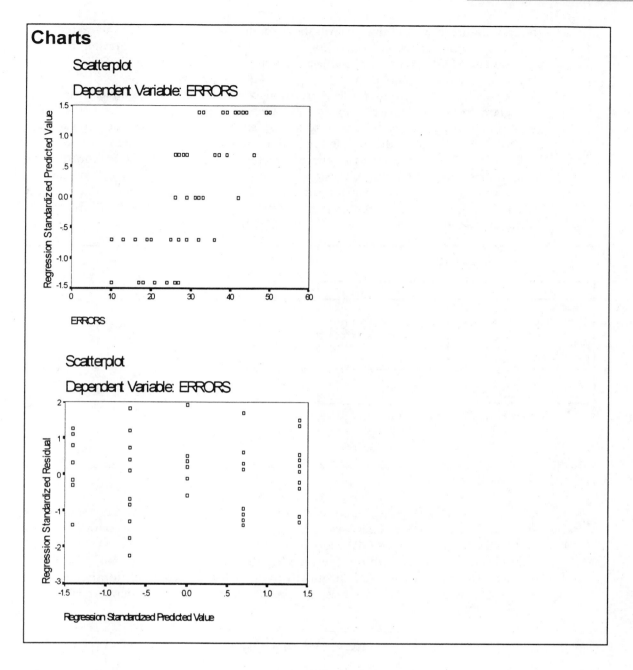

Pearson's Correlation Coefficient

Occasionally we are not interested in predicting one variable on the basis of our knowledge of another variable. We merely wish some index of how closely the two variables
are related to each other. The statistic of choice in such cases is Pearson's product-
moment correlation coefficient, r. As stated in the text, this statistic enjoys a tremen-
dous amount of popularity, though it should probably always be kept in mind that the
underlying model of r is the same as that of regression (which is why the symbol is an
r to begin with), namely that there is a linear, functional relationship between your two
variables. As regression is a more fully elaborated technique, allowing for many nuances
of application, and many diagnostic techniques, it is often to be preferred over r.
However, occasionally you will just want a correlation between two variables without

thought of direction, and thus we will show you how to calculate one. The example we will use is the third data set from Section 9.1 of the textbook, the relationship between IQ scores and GPA scores for 50 subjects. These data can be found in file **DS09_03.dat** and we have labeled the two variables IQ and GPA respectively.

To get SPSS to produce the correlation coefficient is quite simple. Just select **Correlate** from the **Statistics** menu, and then choose **Bivariate...** from the listed options. The resulting dialogue is quite self explanatory. Simply move the two variables you wish to correlate to the middle box, using the arrow button. Note that you could make many pairings of variables and have SPSS test them all. When you have made your selection press the **OK** button, using all the default settings. This will produce a Pearson's r for each of your pairs of variables, and an accompanying p-value, calculated by converting your r to a t-test, with the appropriate degrees of freedom. The results of this analysis follow.

Correlations

		IQ	GPA
Pearson Correlation	IQ	1.000	.534**
	GPA	.534**	1.000
Sig. (2-tailed)	IQ	.	.000
	GPA	.000	.
N	IQ	50	50
	GPA	50	50

**. Correlation is significant at the 0.01 level (2-tailed).

As can be seen, the value of our correlation coefficient is 0.534. You can square this result to get a measure of effect size. $(0.534)^2 = 0.285$, thus we can account for 28.5% of the variation in a persons GPA score by knowing their IQ (and vice versa). The probability values associated with the t-test are less than 0.01 according to our output, and thus we have reason to discredit the null hypothesis that the actual relationship between these scores is 0. We thus have reason to believe that there is a moderate relationship between IQ and GPA.

Review of Concepts

■ Regression analysis is a technique which allows us to generate models of a functional, linear relationship between two (or more) variables. The form of the model is given by the slope and intercept of the straight line that best describes this relationship.

■ One can test the fit of a regression model in (at least) three ways. One can examine the size of effect of the model, given by the model's R-square. One can examine an ANOVA which tests the null hypothesis that our coefficients are all zero. Finally, one can examine the residuals of our model (the predicted values minus our original response values) graphically.

■ To perform a regression analysis in SPSS one uses the **Linear...** item from the **Regression** item on the **Statistics** menu. The dialogue opened has options for selecting our model, choosing output statistics, and requesting plots.

■ Occasionally one is interested in a non-directional (and thus non-predictive) measure of association between two variables. The statistic of choice in this situation is Pearson's correlation coefficient, r. It can be evaluated by squaring it to give a magnitude of effect measure, and also by converting it to a t statistic testing the null-hypothesis that the actual relationship between the two variables is zero.

■ To request a correlation between two variables, you select **Bivariate...** from the **Correlate** item from the Statistics menu. The default options are usually all you will want or need.

Exercises

1. Example 2 in Section 9.1 examined test scores as a function of age. Using the file **DS09_02.dat**, obtain a linear regression summary table, estimates of α and β, and a 95% confidence interval for β.

2. Problem 4 of Section 9.1 investigated the relation between maternal attention and infant attractiveness. Using the file **DS09_09.dat** obtain the Pearson product-moment correlation coefficient between these two variables.

3. Problem 1 of Section 9.2 investigated the relation between memory and marijuana. Using the file **DS09_06.dat**, obtain a linear regression summary table, estimates of α and β and a 95% confidence interval for β.

    Using the files **DS09_13.dat**, **DS09_14.dat**, and **DS09_15.dat**, obtain Pearson product-moment correlation coefficients for the IQs of fraternal twins reared together, identical twins reared apart, and identical twins reared together (Chapter 9, Exercise 2).

7 ANALYSIS OF CONTINGENCY TABLES

Goodness of Fit

In chapter 10 of the Textbook you were introduced to the analysis of categorical data. While we have already used one categorical variable in both independent samples *t*-tests and analysis of variance, both of those procedures had response variables that were continuous. When we have two (or more) variables that are categorical and we wish to test the frequency of subjects in each of those categories, we must rely on other procedures. The simplest of these is the goodness of fit test, which has only one variable, and the frequencies of subjects in the different levels of that variable.

For our example of a one-way table, suitable for a goodness of fit test, we will use one of the examples from Chapter 10 of the text. This is the experiment in which the researchers asks each student in a class to think of a single digit ranging from 0 to 9. The data for this experiment is not on the disk, but as you will see, is easy enough for us to enter directly. To do this we use the data editor. For our example you would create a variable called Digit defined as a numeric variable with no decimal places, and a variable called count which defined similarly (Review Sections 1 and 2 of the Workbook if uncertain about data entry and definition). You then select **Weight Cases** from the **Data** menu. This calls up a dialogue from which you can select a variable in your file to hold frequency counts for each case in your data set. Press the control labeled **Weight cases by...** and then move Count from your list of variables to the adjacent box using the arrow button. Click **OK** to finish. Now SPSS will treat your data set as if their were as many cases of each level of digit as you have stored in the count variable. Now fill in the numbers from the example in the Text book, for example Digit 0 has a Count of 13. This is a handy shortcut compared with creating a file that actually has 13 cases with Digit=0.

To perform our Goodness of Fit test, select **Nonparametric Tests** from the **Statistics** menu. From the list of available alternatives select **Chi Square....** In the resulting dialogue move Digit from your list of variables to the **Test Variable List** box using the arrow button. Then select **OK**, and the default options will produce the output shown on page 145.

The first half of the output is of course our one-way contingency table, listing our observed frequencies and our expected frequencies. The second part gives us our χ^2 of independence. Note that the value of 54.963 (nearly identical to the 55 computed by hand in the Text minus rounding errors) has an associated significance level of .000. To report this result you could say that your obtained $\chi^2(9) = 54.963$, $p<0.001$. We can

NPar Tests

Chi-Square Test

Frequencies

DIGIT

	Observed N	Expected N	Residual
0	13	27.0	-14.0
1	16	27.0	-11.0
2	19	27.0	-8.0
3	51	27.0	24.0
4	26	27.0	-1.0
5	28	27.0	1.0
6	37	27.0	10.0
7	43	27.0	16.0
8	22	27.0	-5.0
9	15	27.0	-12.0
Total	270		

Test Statistics

	DIGIT
Chi-Square[a]	54.963
df	9
Asymp. Sig.	.000

a. 0 cells (.0%) have expected frequencies less than 5. The minimum expected cell frequency is 27.0.

thus say we have discredited the null hypothesis that the frequencies of digits imagined by our subjects were equal.

2 × 2 Contingency Tables

For our example of a 2 × 2 contingency table we will again lift an example directly from the pages of the textbook. In this experiment subjects were asked to state a preference for photos, when the available choices were either a photo of themselves, or a photo of a good friend, when either photo could be displayed in either normal or mirror image orientation. It was predicted that subjects should prefer the normal orientation of their friend's face but the mirror image orientation of their own face, both corresponding to their usual experience. Use the procedures described above to enter the data and weight the cases appropriately. Note that SPSS expects numeric variables for our row and column levels, we can make our table more readable using value labels, however. For our example we have named the row variable Photo and given it values of 1 if the photo is of the subjects' self and 2 if the photo is of a friend. Similarly we have labeled the Column variable Orient and given it a value of 1 for normal orientation and 2 for mirror-image orientation. We then used the **Labels** button on the data definition dialogue (see Section 2) to give Orient its full name of Orientation, and to supply the appropriate value labels for each level of our numerically coded variables. Finally we included a variable called count which we used to enter the obtained frequencies from the contingency table in the textbook.

To perform a χ^2 test of independence on contingency tables with more than one variable you do not use the procedure described above. Instead select **Crosstabs...** from the **Summarize** item on the **Statistics** menu. On the resulting dialogue, move your variable Photo to the **Rows** box and Orient to the **Columns** box. Then select the **Cells** button and place a check mark next to **Expected** as well as **Observed** so that we will have our expected frequency counts printed in our table. After pressing **Continue** select the **Statistics...** button and place a check mark next to **Chi-Square** and **Phi and Cramer's V**. This will produce the tests we wish to examine. After making these selections press **OK**, and you will obtain the output shown on page 147.

First we are presented with a processing summary. This is where we would be informed of any cases that had to be dropped from our analysis due to missing values for either of our variables. Note that due to the way we entered our data using the weighting cases option, it isn't actually possible that we would have a bad case. Next we are presented with our contingency table, which should look quite familiar. The next table gives us the results of our chi-square test. As you can see the value of 12 is significant at $p<.001$, which agrees with the results obtained in the textbook. Finally we have our table containing our value of phi, which is our measure of strength of association, and thus effect size, analogous to Pearson's r. Its absolute value of 0.49 is quite respectable, particularly in comparison to results reported in many studies in the psychological literature.

Larger Contingency Tables

The methodology employed above extends directly to cases with more than 2 rows and/or columns. For an example of a larger contingency table we will use a data file found on your disk, and corresponding to Exercise 2 in Chapter 10. Accordingly the data will be found in the file **DS10_01.dat**. The data are in raw form. In other words, each line of data corresponds to one subject, with the two variables indicating which of the three groups they are members of and which of the two responses they made, respectively. Thus we will not have to weight our cases, SPSS can count the cell frequencies for us. Unfortunately the responses are recorded as a string variable, coded A and R. As stated above SPSS requires numeric variables for cross tabs analysis, so we will have to recode these values to numbers, say 1 and 2 respectively, using the **Recode** option from the **Transform** menu using methods variable. To make the output easier to read, we have chosen to add value labels, described in Section 2, so that SPSS will print an R for responses recoded as 1 and an A for responses recoded as a 2. The analysis follows exactly the same procedure as with the 2×2 contingency table above, and we have made the same selections. The results are on page 148.

The output has the same format as for the 2×2 contingency table. Our interpretation parallels that case, as well. Our χ^2 is once again significant with a value of 10.069 at 2 degrees of freedom. It should be noted that when we are looking at tables with more than 2 rows and columns, however, that we have to make an adjustment to our

Crosstabs

Case Processing Summary

| | Cases | | | | | |
| | Valid | | Missing | | Total | |
	N	Percent	N	Percent	N	Percent
PHOTO * orientation	50	100.0%	0	.0%	50	100.0%

PHOTO * orientation Crosstabulation

| | | | orientation | | Total |
			normal	mirror image	
PHOTO	self	Count	9	16	25
		Expected Count	15.0	10.0	25.0
	friend	Count	21	4	25
		Expected Count	15.0	10.0	25.0
Total		Count	30	20	50
		Expected Count	30.0	20.0	50.0

Chi-Square Tests

	Value	df	Asymp. Sig. (2-tailed)	Exact Sig. (2-tailed)	Exact Sig. (1-tailed)
Pearson Chi-Square	12.000[b]	1	.001		
Continuity Correction[a]	10.083	1	.001		
Likelihood Ratio	12.647	1	.000		
Fisher's Exact Test				.001	.001
Linear-by-Linear Association	11.760	1	.001		
N of Valid Cases	50				

a. Computed only for a 2x2 table

b. 0 cells (.0%) have expected count less than 5. The minimum expected count is 10.00.

Symmetric Measures

		Value	Approx. Sig.
Nominal by Nominal	Phi	-.490	.001
	Cramer's V	.490	.001
N of Valid Cases		50	

Crosstabs

Case Processing Summary

	Cases					
	Valid		Missing		Total	
	N	Percent	N	Percent	N	Percent
GROUP * RESP	108	100.0%	0	.0%	108	100.0%

GROUP * RESP Crosstabulation

			RESP		Total
			R	A	
GROUP	1	Count	28	8	36
		Expected Count	23.3	12.7	36.0
	2	Count	16	20	36
		Expected Count	23.3	12.7	36.0
	3	Count	26	10	36
		Expected Count	23.3	12.7	36.0
Total		Count	70	38	108
		Expected Count	70.0	38.0	108.0

Chi-Square Tests

	Value	df	Asymp. Sig. (2-tailed)
Pearson Chi-Square	10.069[a]	2	.007
Likelihood Ratio	9.954	2	.007
N of Valid Cases	108		

a. 0 cells (.0%) have expected count less than 5. The minimum expected count is 12.67.

Symmetric Measures

		Value	Approx. Sig.
Nominal by Nominal	Phi	.305	.007
	Cramer's V	.305	.007
N of Valid Cases		108	

calculation of phi. As stated in the text book, this adjustment was proposed by Cramér, and the statistic is thus commonly known as Cramér's phi, which is somewhat inexplicably rendered as Cramér's V in the SPSS output. Presumably it was felt that a V was the closest ascii character available to a phi, but that still doesn't explain why the statistic wasn't simply labeled Cramér's phi. Whatever the explanation, we have a value of 0.305 for this statistic which is still of moderate size, though less than that of our previous example.

Review of Concepts

■ When we wish to analyze data in the form of frequency counts, the methodology involves the construction of contingency tables and the subsequent testing of observed frequencies against the expected values of frequencies under the null hypothesis of independence.

■ We can enter our frequency counts directly using the **Weight Cases** item from the **Data** menu, or we can enter raw data and have SPSS count the frequencies itself.

■ In SPSS the tests of independence are conducted using two different procedures.

■ We use the **Chi Square** option from the **Nonparametric Tests** item from the **Statistics** menu for simple goodness of fit tests.

■ For all contingency tables with more than one variable, you use the **Crosstabs...** item from the Statistics menu. You can add the expected frequencies from the **Cells** button and request a chi square test and phi coefficient from the **Statistics** button.

■ The phi coefficient and Cramér's phi (labeled V) can be interpreted as measures of effect size analogous to Pearson's r.

Exercises

Answer the questions in Exercises 3 and 4 of Chapter 10 using SPSS.

Introduction to the Use of SAS

1   Introduction to SAS: Reading in Data from Disk

2   Examining and Describing Data

3   Obtaining Standard Errors and Confidence Intervals

4   Analysis of Variance for Independent Groups

5   Matched Pairs and Within-Subjects Designs

6   Regression and Correlation

7   Analysis of Contingency Tables

1 INTRODUCTION TO SAS: READING IN DATA FROM DISK

What Is SAS?

SAS is a collection of programs originally developed for statistical analysis but now encompassing many database and economic forecasting capabilities as well. The programs are available for most platforms of computers ranging from large institutional mainframes, to a home PC or Macintosh. The SAS Institute estimates that SAS is used at over 3 million sites worldwide, including universities, hospitals, governmental laboratories and large businesses.

The SAS Institute in Cary, North Carolina, provide excellent customer service, normally through a representative at each large institute where SAS is available. Your university probably has a SAS representative, though we anticipate that for most of you, your help will come from your course instructor and teaching assistants. Because of the diversity of platforms SAS is available for, it would be impossible to include instructions here on each operating system (like UNIX, DOS or Windows 95) that SAS might by running under. We therefore assume that you have some working knowledge of how to operate your computer (things like turning it on, logging in, finding files in directories, etc.). Likewise, we omit any particular short-cuts your version of SAS may have (like windows and menus) in order to make this workbook as generally useful as we can. We rely on your instructor(s) to provide additional information about the computers at your site and any special features of your version of SAS they wish you to use.

The programs in this workbook were written using SAS version 6.11 and tested on both an IBM-compatible PC running Windows 95 and a Sun Workstation running UNIX. These programs should be usable on virtually any version of SAS later than version 5 and on virtually any computer platform running any operating system.

What Is a SAS Program?

SAS is a programming language, like Basic or C, which provides your computer with a list of instructions to be carried out in a given order. Writing a program in SAS is a lot like writing a list of directions for a friend whom you want to run some errands for you. It can be broken down into two parts: the Data Step and the procedures or Procs. The Data Step tells SAS how to read your data, whether you type it in as part of the SAS program or have the program read it from a separate data file. The Proc steps are commands that have SAS do things to your data, like calculate means or perform a t-test.

In total, a given run of a SAS program involves three files, one produced by you and two produced by SAS. The file produced by you is the SAS code, which is just the text of your program as you typed it. You should usually usually give this file a meaningful **filename** and then an extension like **.sas** to remind you that it contains instructions for SAS. When you run this program you are telling SAS to execute the instructions in the code file, one by one, in the order they are written. On most large computers running UNIX or a similar operating system, you execute a code file by typing sas and then your **filename**. If you are using SAS for Windows or DOS, you can execute a code file as soon as you've written it by using the **Submit** command from the **Locals** menu or by pressing the **F8** function key while the window with the code file in it is active.

Once you have run your program, SAS creates two new files for you: a Listing file and a Log file. Both of these files will have the same filename as your code file but will have the extensions **.lst** and **.log**, respectively. The listing file contains the output of the various instructions you gave to SAS. Any statistics you requested will be here, as will any charts or graphs. The log file contains a line-by-line copy of your program code, interspersed with messages from SAS about the success or failure of each instruction. It is in this file that you will discover why your output is different from what you had expected, probably because of typing errors in your commands, improper instructions, missing semicolons, etc.

Throughout the SAS sections of this workbook we provide you with sample programs that you can imitate or even directly copy. These can serve as frameworks for your own analyses. We utilize a number of conventions when we present these programs in order to make them consistent with one another and easy for you to interpret.

> ■   Each program is enclosed in a box, like this one.

- ■   We write the names of any special SAS commands in all upper-case letters, as we have been doing for the name SAS.

- ■   We use lower case with initial capitals for any variable names and lower case for optional statements that modify how the program interprets the commands.

- ■   The output generated by each program is enclosed in a box, as well.

Where possible we present output that is identical to what you will receive in your own output file when you run the program. Occasionally, in the interest of saving space, we reduce the size of graphs and charts by about 50%.

How to Write SAS Programs

Basically you can use any text editor or word processor to write your SAS code file, as long as you save the result as plain text (most word processors allow this). You write your instructions out, and follow each with a semicolon (;). This tells SAS that it has reached the end of one instruction and should carry it out before proceeding to read the next instruction. We will show you a simple example of a SAS program that reads in three variables from each line of data and then prints out the data set, and then discuss each line of the program to let you know what's going on.

Example 1

```
*This is example 1;
OPTIONS ls=80 ps=60;

DATA new;

     INPUT Gender $ Wage Hours;

LINES;
M    21    15
F    25    17
M    23    16
RUN;

PROC PRINT data=new;
     var Gender Wage Hours;
RUN;
```

Let's look at what this program is asking SAS to do. Remember SAS reads each program, one line at a time, stopping to execute each instruction when it reaches a semicolon. The fact that we've indented some of the lines makes no difference to SAS. SAS ignores blank spaces, except in lines of data (see below). The indentations are just a programming convention to make our program more readable to us.

The very first line begins with an asterisk (*). A line beginning with an asterisk is called a comment line, and will be ignored by SAS. As far as SAS is concerned everything between an asterisk and the semicolon at the end of that line, does not exist. This is handy as it gives you a way to write comments or reminders into your program without them interfering with the way your program runs. When you come back to use a program you haven't looked at in a long time, you may no longer be sure of what it does. If you left some comment lines explaining when the program was written, and what it was for, you can save yourself a lot of headaches.

The second line contains the SAS command OPTIONS. The OPTIONS command lets you control such features of the output from your program as the length of each page, the length of each line, whether SAS should print the date at the top of each page, etc. In this case we followed the command name with two command line options: ls=80 and ps=60. The first of these tells SAS to use 80 characters per line in the output file (line size = 80) and the second tells SAS to print the output file using pages of 60 lines length (page size = 60). You don't need to include an OPTIONS command in every SAS program, but it often helps to make your output more readable.

The next line begins the Data step, by invoking the DATA command and creating a temporary data set called new. You can use any name for your temporary data set, as long as it begins with a letter, contains only letters or numbers, and is eight characters long or shorter. You can see that our data set name (new) meets all these criteria. You should choose a name that is meaningful to you.

The next line is the INPUT command. It tells SAS how to read each line of data. SAS is very flexible in how it reads lines of data. We are using one of three different input styles, called list input. With list input (which is the easiest) SAS reads each variable name, and assumes that there will be a corresponding value for each variable on each line. As SAS reads each line of data, it will interpret any blank spaces as breaks between the variables. The rest of the INPUT line contains a list of our variable names. They must all obey the same criteria as data set names. SAS recognized two kinds of data, strings and numeric. Strings are any data which include letters, while numeric variables contain only numbers. When you wish SAS to read in a string variable you have to follow that variable name with a dollar sign ($) which is a symbol generally used for strings in computer programming. Thus with this INPUT statement, SAS has been told that each line of data contains three variables. The first is a string variable called Gender, the second is a numeric variable called Wage and the third is a numeric variable called Hours. Always try to give your variables meaningful names, rather than var1, var2, or A, B, etc.

The next line in the data step, LINES, tells SAS that the data are starting. The commands CARDS and DATALINES are synonymous with LINES. SAS will accept any of these three, as the start of the data. SAS will now read each line of data looking for a string variable, followed by two numeric variables. Note that if there where a fourth column of data, after these three, SAS would ignore it. Since we told it that there were only three variables on each line, once it has read three variables it proceeds to the next line.

On a related note imagine that the Wage variable was missing from the data file for the third observation. The line would now look like this:

F   17

SAS would be reading the line looking for one string variable followed by two numeric variables, separated by spaces. Thus it would read the Gender of F correctly, and then skip over any blank spaces until it reached its first numeric variable, in this case a 17 which it would dutifully record as the value for Wage, and then continue to read the line looking for a number to record for Hours. When reading data in list format, SAS has no way of knowing that a blank might mean data is missing. For this reason it is very import to check your data files to make sure there are no missing values, and if there are, to place something in the blank space to tell SAS there's a missing value. Usually you would insert a period (.) in the place where the missing value would be. Doing that in our example would change the line to

F . 17

and thus SAS would read the value F for Gender, register the value of Wage as missing, and the value of Hours as 17. The other solution to this problem is to use a more complicated INPUT line then the list input we've been using thus far, but that would take us beyond the scope of this chapter.

The data step ends with the first semicolon following that lines of data. That is why we don't put a semicolon at the end of each line of data. If we did, SAS would stop inputting data when it reached the end of the first line. Because it hasn't reached a semicolon, SAS will keep reading in lines of data, until it does reach one. We have inserted a RUN; command here, although it isn't strictly necessary it is good practice, and it will remind you to put a semicolon in after all the data lines. The data step is now finished. We can now request any procedures we wish to have performed on these data.

The next line is a command for a procedure. All procedures begin with the word PROC (for PROCedure, of course) followed by the name of the procedure to be carried out and end with the command RUN. Here we have requested the PRINT procedure, which tells SAS to list our data set for us. We have included a command line option to tell SAS which data set we want listed (remember we called our data set "new"). If we hadn't included this option SAS would have simply printed the most recent data set it had worked with. In this case, it wouldn't have made any difference, but once you start writing longer, more complicated programs, some of which may have data steps intermingled with procedure steps, it could make a difference; so it's to your advantage to begin some good programming habits now.

The lines in between a PROC command line and the RUN that ends a procedure are used to give optional statements, which modify how the procedure is carried out. Sometimes these statements are required for SAS to know how to properly perform a procedure. When that is the case we print them in upper case, just as we're doing with all required instructions. Here we have just included an optional statement, the var statement common to many procedures. The var statement just tells SAS which variables we want the procedure carried out on. We have listed all three variables in the var statement, and thus SAS will print all three variables in the listing. Note that the var statement is optional, when you don't supply a list of variables, SAS will just print (or execute another procedure) on all the variables in your data set. Again, in this case it wouldn't have made a difference, but I'm sure you can imagine situations in which it would. Also note, that when we refer to the variable Gender we don't have to follow it with a dollar sign ($). That symbol is only needed in the INPUT line to tell SAS that the variable is a string, thereafter whenever you refer to the variable you just use the name you've given it.

Here is the output of the program in example 1.

```
The SAS System                           1
                                        17:12 Monday, July 21, 1997

            OBS     GENDER    WAGE    HOURS

             1        M        21      15
             2        F        25      17
             3        M        23      16
```

Note that SAS prints our listing using the options we defined at the top of our program (and thus we had to reduce the font size to fit all 80 columns on the page of this workbook). At the top of each page of output you get the SAS system message as well as a page number. Following this SAS prints the date and time (though we could have suppressed this by adding nodate to our list of options). There follows our list of variables, centered under the names we gave them. Double check to make sure that SAS read in everything correctly. Note that SAS adds a variable of its own called OBS (for OBServation) which is just a counter to let you (and SAS) now which line of input each line of output came from.

Here is the log file of this program.

```
NOTE: Copyright (c) 1989-1996 by SAS Institute Inc., Cary, NC, USA.
NOTE: SAS (r) Proprietary Software Release 6.11  TS040
      Licensed to UNIVERSITY OF TORONTO COMPUTING & COMMUNICATIONS, Site 0008987001.


1     *This is example 1;
2     OPTIONS ls=80 ps=60;
3
4     DATA new;
5
6          INPUT Gender $ Wage Hours;
7
8     LINES;

NOTE: The data set WORK.NEW has 3 observations and 3 variables.
NOTE: The DATA statement used 2.79 seconds.


12    RUN;
13
14    PROC PRINT data=new;
15         var Gender Wage Hours;
16    RUN;

NOTE: The PROCEDURE PRINT used 1.05 seconds.
```

As you see, SAS carefully read through our instructions and carried them out, one-by-one, until the end of the program file. The notes inserted by SAS indicate that there were no problems with this program. Pay special attention to the note that tells you how many observations and how many variables there are in your data set. Often when you get unexpected results from your program, it may be because you and SAS disagree about these numbers (usually because some data was typed incorrectly or a semicolon was inserted where it didn't belong) and thus this note is an important guide to where in your data step you may have gone wrong.

Reading Data from Files

The example above illustrated how data could be entered into SAS for further analysis by typing it into the data step directly, between the command LINES and a final semi-colon. You should feel free to type in any data from your text book or work book to try them out; however there is an easier way of handling the data from the examples in the textbook and workbook All of the data sets used in either book are available as text (ASCII) files either from your instructor or by downloading the files from the book's web site.

Data sets from the textbook have filenames corresponding to their designations in the textbook itself. Thus Data Set 2.1 in the textbook adopts the filename **DS02_01.dat**; Data Set 2.4 has the file name **DS02_04.dat**; data Set 3.1 has the filename **DS03_01.dat**; and so forth. The leading zeros (01, 02, etc.) have been inserted so that when the files are listed on your screen, they will be correctly ordered. Without the leading zeros, most software when reading filenames place the number 11 before the number 2. However 02 will be placed before 11. The filename extension ".dat" has been added because both SPSS and SAS look for this extension as designating a data set. Data sets from the workbook have file names denoting one of the three research areas used in the workbook exercises. Data sets for the memory area are named MEM_01.dat, MEM_02.dat, etc. Data sets for the child behavior area are named CHILD_1.dat, CHILD_2.dat, etc. Data sets for the exercises involving test construction and evaluation are named TEST_1.dat, TEST_2.dat, etc. For convenient reference, all data sets are listed in Appendix B of the workbook.

To use one of these data files you will first have to copy it to a directory on the computer on which you intend to use SAS. Ask your instructor if you are uncertain about how to do this. For now, we will assume you have copied all of data files from the disk to a directory on your computer called datasets on the hard drive (C:) of your IBM PC. Now assuming that you are trying to get SAS to read the data in the first data set mentioned above, and that you are working on an IBM PC, you could write a program as follows.

Example 2

```
OPTIONS ls=80 ps=56 nodate;

DATA children;

        INFILE 'c:\datasets\DS02_04.dat';
        INPUT Agegroup Vocab;

RUN;
```

As you can see, there are two dramatic differences between this data step and the one we used in the first example. First there is the INFILE command. This command tells SAS where to find the file with the data. Second you will notice the absence of a LINES (or CARDS or DATALINES) command. This is no longer necessary, since SAS will be reading our data directly from a file. The name for the temporary data set we are creating, and for the variables were chosen to be meaningful, on the basis of the data involved. Take a look at the third data set used in Section 2.1 of the textbook to confirm the meaningfulness of our choices.

As we stated above, we have assumed that you have loaded all the data and that they reside in a directory called datasets, on the hard drive (C:) of your IBM PC. If you have the data saved elsewhere, or are using an operating system other than DOS or Windows, you will have to adjust the command accordingly. The important point to remember is to place the filename you wish to load, along with the pathway of directories to where the file is stored, between the single quotation marks following the INFILE command. Thus to load this file off a diskette, you would use the INFILE command

```
INFILE 'a:\datasets\DS02_04.dat';
```

This assumes that you have a floppy disk drive of the correct size labeled a: on your system, and that you have inserted the diskette into that drive. On a UNIX-based machine you might well have a much more intricate pathname.

Additional Data Step Commands

You can use the data step to do more than just load data. You can use other commands, designed to manipulate the data you have loaded to transform variables, or create new ones, using many common mathematical functions. For example, note the new variables created in the following example.

Example 3

```
OPTIONS ls=80 ps=56 nodate;

DATA weather;
        INFILE 'c:\weather.dat';
        INPUT TempF Ampress Pmpress Proprain;
            TempC = ((TempF * 5) / 9) - 32;
            Avgpress = (Ampress + Pmpress) / 2;
RUN;
```

In this program we are loading a fictional data file called weather.dat (you don't have this file, so don't look for it on the disk). The file contains four variables: the temperature in degrees Fahrenheit, the air pressure measured at 12:00 a.m., the air pressure measured at 12:00 p.m., and the proportion of the day it rained. We have created two new variables using transformations of the data loaded: the temperature in degrees Celsius, and the average air pressure for the day. The first two variables are created using mathematical operators on the variables as we load them. The legal mathematical operators in SAS follow.

Operators used	What they do
x + y	Addition of x and y
x − y	Subtraction of y from x
x * y	Multiplication of x and y
x / y	Division of x by y
x ^ y	Raising of x to the power of y

Additionally, there are a number of mathematical functions which can be used, such as the arcsine and square root functions used in the above example. In general, to use one of these functions you first write the SAS command for the function (usually an abbreviation of the function name) and then place the name variable you want transformed in parentheses, this is called the argument of the function. The SQRT(x) means "return the square root of x". A list of some of the more common functions that can be used in this way follows:

Function called	What it does
LOG(x)	Return the natural logarithm of x
LOG10(x)	Return the base 10 logarithm of x
SIN(x)	Return the sine of x (in radians)
COS(x)	Return the cosine of x
TAN(x)	Return the tangent of x
ARSIN(x)	Return the arcsine (inverse sine) of x
SQRT(x)	Return the square root of x
FLOOR(x)	Return x with any decimal fractions removed.
ROUND(x, y)	Round off the value of x to the nearest y

Note that the ROUND function takes two arguments, the variable x which is what you want rounded off, and y which is what you want it rounded to. For example, to round off the temperature in degrees Celsius to the nearest $1/10^{th}$ of a degree you could use the command ROUND(tempC, 0.1). Similarly to round the temperature off to the nearest 10 degrees you could write ROUND(tempC, 10). The FLOOR function differs from rounding in that it only slices off anything after the decimal place, without altering anything before. For example FLOOR(3.2) = 3, but so does FLOOR(3.9).

As you can see, by combining various arithmetic operators, mathematical functions, and other data step commands we will discuss later in the book, you can exercise a great deal of control over you data with SAS. The ability to manipulate data in almost any form is considered one of the best features of SAS.

One Final Note

Until this point we have only considered data steps that take place before any procedures are called. That is the norm for SAS programming, and certainly will serve for all the examples in this workbook. It should be noted however, that there are many real world situations in which one might want to modify a data set after having performed a procedure on it, and thus a SAS program could conceivably contain several data steps, intermingled with procedures. We have also only considered list input, but you should be aware that SAS can be directed to input data in two other ways: columnar input, and infile formats. Finally, we have only considered data steps which create a temporary SAS data set. A temporary data set is stored in the memory of your computer and lasts only as long as your current SAS session. The next time you run your program the data step creates a new temporary data set from scratch. One might well appreciate that after lengthy processing of data it would be useful to save a SAS data set as a permanent file, on your computer's disk. SAS contains procedures enabling you to do so, but any discussion of these procedures would just complicate our programs for now. Should you wish to learn more about any of these advanced SAS programming issues we direct you to any of the excellent books available on SAS.

Review of Concepts

■ Using SAS to analyze your data involves writing a program in a code file, which is a list of instructions telling SAS what to do. Each SAS command must end with a semicolon (;).

■ SAS responds to these instructions by producing a Listing file of your output, and a Log file of its comments and error messages.

■ A SAS program has two basic parts, a Data Step, which begins with the DATA command, and one or more Procedures, which begin with the command PROC and end with a RUN command.

■ The INPUT command tells SAS to read each line of data looking for values to assign to the variables you have named.

■ There are string variables, whose names are followed by a $ in the INPUT line, and numeric variables.

■ Missing values should be replaced with a period (.) when reading data in list format.

■ Data can be read directly from the program by placing it between the command LINES and a ; at the end of the data step.

■ Data can also be read from a text file using the INFILE command.

■ The procedure PRINT, lists our data set.

■ Data step commands can also be used to create new variables or transform existing variables using arithmetic operators and mathematical functions.

Exercises

The best way to learn any new procedural skill is by trial and error. We therefore encourage you to try loading some data sets, and then printing the results. Don't worry, using the tools we've shown you in this chapter and as long as you let SAS choose the extensions for any files you save, you can't permanently damage any of the data sets on your disk.

The exercises at the end of this and subsequent sections are intended to get you through the initial stages of becoming familiar with the use of SAS to perform the procedures described in that section. There are many more data sets from the textbook and the Workbook on which you can practice.

As a starting point, enter the Data Sets 2.1 through 2.4 either from the keyboard or from the ASCII files **DS02_01.dat**, **DS02_02.dat**, etc. These files will be used in the next section. Suggested variable names are found in the listing of data sets in Appendix B of this workbook. Note that Data Sets 2.1 and 2.2 are non-numerical and the variables must be declared "string." Continue entering data until you are thoroughly familiar with the procedure.

2 EXAMINING AND DESCRIBING DATA

Frequencies

Chapter 2 of the text book begins with the notion of frequencies and methods for tabulating and graphically representing them. In SAS we will use the procedures FREQ to obtain tables of frequencies, and CHART to produce bar charts and histograms. Additionally, in this section, we will examine the procedures MEANS and UNIVARIATE for producing numerical descriptions of our data, as well as Stem-and-leaf and Box-plots, via the latter command. Finally we will discuss some of SAS's high resolution plotting capabilities using the procedures GCHART for high resolution bar charts and histograms.

The following program produces frequency counts and a bar chart for the Infant Response Data stored in the file **DS02_01.dat**. These data corresponds to the first example in Chapter 2 of the textbook.

```
Options ls=80 ps=60;

Data Infants;
        Infile '..\data\DS02_01.dat';
        Input Response $;
Run;

Proc Freq;
        Tables Response;
Run;

Proc Chart;
        Vbar Response;
Run;
```

Refer to Section 1 of the workbook for details of data entry. The output of these procedures is as follows.

```
The SAS System                              1
                                    17:29 Wednesday, March 26, 1997

                                   Cumulative  Cumulative
        RESPONSE   Frequency   Percent  Frequency   Percent
        ƒƒƒƒƒƒƒƒƒƒƒƒƒƒƒƒƒƒƒƒƒƒƒƒƒƒƒƒƒƒƒƒƒƒƒƒƒƒƒƒƒƒƒƒƒƒƒƒƒƒ
        FM            3       11.5        3        11.5
        HO            6       23.1        9        34.6
        LP            4       15.4       13        50.0
        MO            5       19.2       18        69.2
        TP            8       30.8       26       100.0
```

```
                    The SAS System                          2
                                    17:29 Wednesday, March 26, 1997
```

Frequency

By substituting Proc GCHART for CHART in the above program we would obtain the following high-resolution figure.

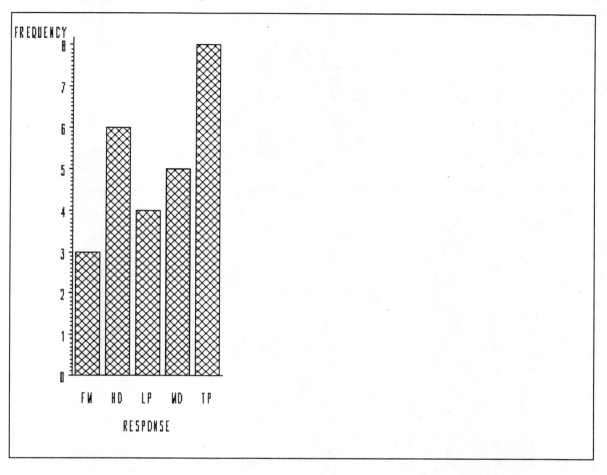

Performing the same operation on data from file **DS02_02.dat**, produces the following frequency table and figure.

```
The SAS System                            6
                                          17:29 Wednesday, March 26, 1997

                                    Cumulative   Cumulative
          BRAND  Frequency  Percent  Frequency    Percent
          ffffffffffffffffffffffffffffffffffffffffffffffffffffff
          A          6       12.0        6          12.0
          B         11       22.0       17          34.0
          C         18       36.0       35          70.0
          D         10       20.0       45          90.0
          E          2        4.0       47          94.0
          F          3        6.0       50         100.0
```

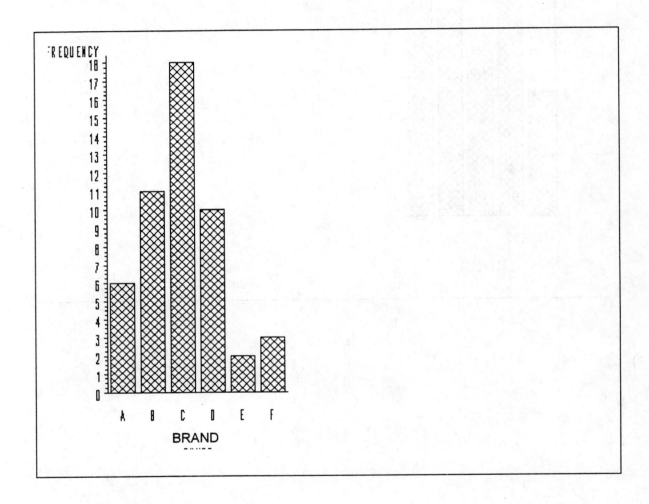

By adding an option to the Vbar statement in Proc Chart we can change the chart from one of raw frequencies to one of percentages.

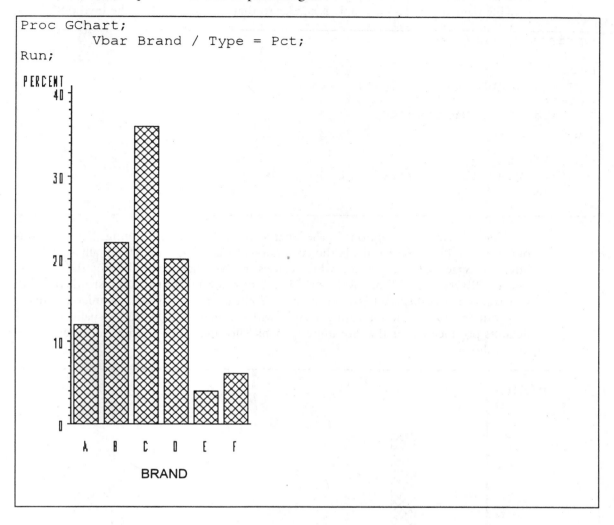

```
Proc GChart;
        Vbar Brand / Type = Pct;
Run;
```

Use of the "Type=" option in our charts allows us to construct charts with frequency counts "Type=Freq" (the default, if no type is specified), percentages (as above) as well as other useful statistics, as follows.

Type = CFREQ	Cumulative Frequencies
Type = CPCT	Cumulative Percentages
Type = SUM	Totals
Type = MEAN	Means

Note that we do not have proportion as an option, and thus we cannot exactly recreate Figure 2.2 from the textbook. This should not be cause for concern, once one remembers that percentages are the same as proportions, simply rescaled to be out of 100 rather than 1.

To examine frequencies of continuous data, we normally use a histogram rather than a bar chart. The following example uses data taken from **DS02_03.dat**, with the variable named IQ. This example is similar to that in Figure 2.4 of the textbook.

```
Options ls=80 ps=60;

Data IQ;
        Input IQ;
Lines;
>>>Data Lines Omitted<<<
Run;

Proc GChart;
        Vbar IQ / Levels = 10 Space = 0;
     Run;
```

The two statement options in the Vbar statement are worth noting. Levels = 10 is our way of telling SAS to divide the continuous variable IQ into 10 equally sized class intervals. Space = 0 is an option which causes the bar chart printed as the default output for PROC CHART or PROC GCHART to print as a histogram instead. In other words it requests 0 space between the bars. You can try setting this parameter yourself to produce more or less pleasing results with ordinary bar charts. Usually the SAS defaults produce charts that are quite readable, however. The output from this procedure follows.

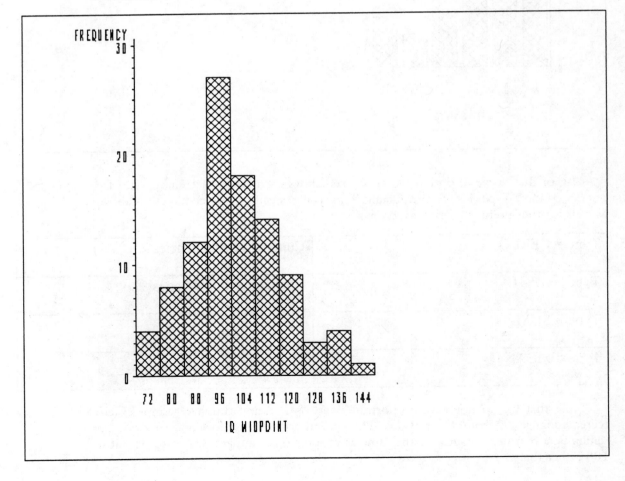

Note that the axis label is not IQ but rather IQ midpoint. The numbers represent the mid points of those 10 equally sized bins created by our Levels=10 statement option above. The example in the textbook, however, had more reasonable lables for the bins. One way to achieve this in SAS is to use a different set of options than we used above.

```
Proc GChart;
        Vbar IQ / Midpoints = 65 to 145 by 10 Space = 0;
Run;
```

This will create the following histogram, with bars approximately identical to those of Figure 2.4.

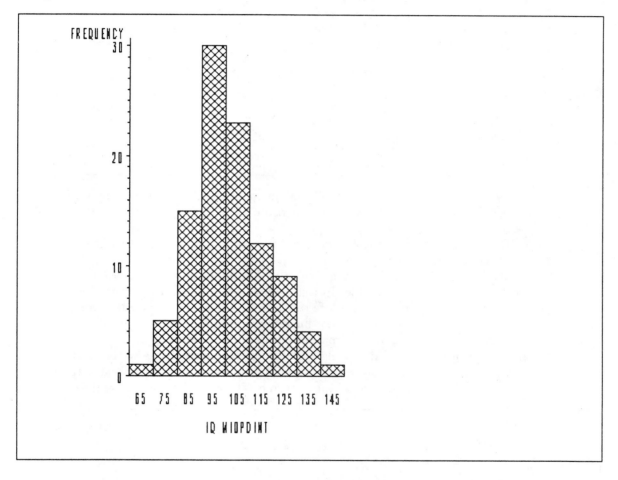

In order to recreate Figure 2.5 we need only change the number of levels in the first example to 20 or the width of the levels from 10 to 5 in the second example. In order to create a table of frequencies with the data divided into class intervals such as Table 2.5 we need to invoke the SAS procedure FORMAT. This will change the values of our IQ variable, recoding them with the values of the midpoints for the class intervals we require. Formatting a variable consists of two parts. The definition of the format, using PROC FORMAT before the data step, and the application of the format to a variable using the Format statement in either the data step or a procedure. The following example illustrates the creation of a new format for IQ, and its application to the IQ variable during the FREQ procedure.

```
Options ls=80 ps=60;

Proc Format;
        Value IQfmt      60-69=65
                         70-79=75
                         80-89=85
                         90-99=95
                         100-109=105
                         110-119=115
                         120-129=125
                         130-139=135
                         140-149=145;

Data IQ;
        Input IQ;
Lines;
>>>Data Omitted<<<
Run;

Proc Freq;
        Format IQ IQfmt.;
        Tables IQ;
Run;
```

Note that the PROC FORMAT statement before the Data Step only creates the format, it doesn't apply it to IQ yet. The name of the format, in this case IQfmt is then specified in the Value statement. IQfmt is a good name because it is a format we will be applying to IQ and thus the chosen name is memorable. Note however that any name could have been applied.

Adding the Format statement to PROC FREQ, or any other procedure, applies a predefined format to a variable. In this case we are applying the format we just created—IQfmt, to the variable IQ. Note that when applying a format, the name of the format ends in a period(.). You do not include the period when defining the format, just when applying it. Using the Format statement within a procedure such as this, formats the variable for that procedure only. It does not change the values of the variable in our data set. In order to format a variable for all procedures we would include the Format statement after the input line of our data step. The result of this program follows.

```
The SAS System                                    3
                             09:20 Thursday, March 27, 1997

                          Cumulative  Cumulative
    IQ   Frequency   Percent  Frequency   Percent
  ƒƒƒƒƒƒƒƒƒƒƒƒƒƒƒƒƒƒƒƒƒƒƒƒƒƒƒƒƒƒƒƒƒƒƒƒƒƒƒƒƒƒƒƒƒƒƒ
    65       1         1.0        1        1.0
    75       5         5.0        6        6.0
    85      15        15.0       21       21.0
    95      30        30.0       51       51.0
   105      23        23.0       74       74.0
   115      12        12.0       86       86.0
   125       9         9.0       95       95.0
   135       4         4.0       99       99.0
   145       1         1.0      100      100.0
```

Note that reformatting data into new numerical values creates an inherent risk that we might forget that this is altered data and begin using the above formatted numbers in future calculations. For this reason it is wise to only invoke formats during the procedure in which we need the formatted data, and to leave the raw data otherwise unchanged.

Using PROC UNIVARIATE

To obtain the stemleaf plots and boxplots discussed later in chapter 2 of the text we will have to use PROC UNIVARIATE. This procedure is one of the most versatile and useful that SAS has to offer, allowing for a wide range of numerical and graphical descriptions of data. The syntax is fairly straightforward. Using our IQ data as above, we have

```
Proc Univariate plot;
        Var IQ;
Run;
```

The word "plot" in the above command is a procedure option, following the name of the procedure (UNIVARIATE in this case). This option requests the plots we are interested in. Note the difference between the syntax of this procedure option and the statement options used in PROC GCHART above. Statement options are seperated from their statement with a slash "/", while procedure options just follow the procedure name, without any seperator. The Var statement lists the variables we wish the UNIVARIATE procedure to examine. The default, if no Var statement were included, would be to perform PROC UNIVARIATE on every numerical variable in the data set. In this case it would be no different, as IQ is the only variable in **DS02_03.dat**. It is often wise, to save computing time, printer paper, and just the annoyance of having UNIVARIATE operate on irrelevant variables (e.g., Subject Numbers) to always specify the variables you are interested in. The output from this procedure appears on pages 172–173.

As you can see, UNIVARIATE gives you a lot of output. Don't let the density of this printout discourage you. Let us examine some of the relevant sections. First of all the variable processed is listed. In this case the variable is IQ. It is a good idea to get in the habit of checking this variable name whenever you pick up a printout from this procedure, just to make sure you are examining the variable you think you are.

Next, we have a section labeled moments. Without getting into the mathematics of what precisely a moment is, there are some numerical descriptors here you should be familiar with. According to the output we have an N of 100, a Mean of 101.17, a Standard Deviation of 15.3248, and a Variance of 234.8496. Just as its a good idea to check the variable name, it is always a good idea to check the value of N, to make sure that you and SAS agree on how many subjects are in your sample. If the number of N seems wrong, you may have made an error in data entry, and should double check the data step of your SAS code. One other moment worth mentioning in passing is a Skewness of 0.43538. You were introduced to skew as a visual cue you could look for in plots of your data. The result presented here are simply numerical summary calculated directly from the data. A large number for Skewness reflects a problem which should also be apparent on visual examination. Negative and positive values of Skewness reflect negative and positive skewing, respectively. None of the other moments listed, need concern you at this time, they all have limited use, for quite specific applications.

The next section of output is labeled Quantiles. Quantiles is a general term to refer to values such as quartiles, deciles, etc. Again some of the descriptors should be familiar to you. For example the 50% quantile is also labeled Med for median. Other quantiles listed are also useful such as Q1 and Q3 which give you the interquartile range, helpfully calculated for you and displayed below as Q1-Q3 range. Additionally, this section lists the minimum value, the maximum value, the overall range of your data, and the mode.

```
The SAS System                              1
                                    09:20 Thursday, March 27, 1997

                       Univariate Procedure

Variable=IQ

                            Moments

         N               100    Sum Wgts          100
         Mean          101.17   Sum            10117
         Std Dev      15.3248   Variance     234.8496
         Skewness     0.43538   Kurtosis      -0.0074
         USS          1046787   CSS          23250.11
         CV          15.14758   Std Mean      1.53248
         T:Mean=0    66.01716   Pr>|T|         0.0001
         Num ^= 0         100   Num > 0          100
         M(Sign)           50   Pr>=|M|        0.0001
         Sgn Rank        2525   Pr>=|S|        0.0001


                       Quantiles(Def=5)

         100% Max         142   99%            139.5
          75% Q3          111   95%              130
          50% Med          98   90%              123
          25% Q1           92   10%               82
           0% Min          69    5%               79
                                 1%               70

         Range             73
         Q3-Q1             19
         Mode              96


                            Extremes

         Lowest     Obs      Highest     Obs
             69(       4)       133(       44)
             71(      67)       135(       95)
             74(      11)       137(       16)
             75(      36)       137(       31)
             79(      45)       142(       56)


                       The SAS System                         2
                                    09:20 Thursday, March 27, 1997

                       Univariate Procedure

Variable=IQ

    Stem Leaf                        #            Boxplot
      14 2                           1               0
      13 577                         3               |
      13 3                           1               |
      12 7                           1               |
      12 01233344                    8               |
      11 789                         3               |
      11 113334444                   9            +-----+
      10 566788899                   9            |     |
      10 00011223334444             14            |  +  |
       9 556666666677777788         18            *-----*
       9 111222233333               12            +-----+
       8 5788889                     7               |
       8 00122344                    8               |
       7 599                         3               |
       7 14                          2               |
       6 9                           1               |
         ----+----+----+----+
    Multiply Stem.Leaf by 10**+1
```

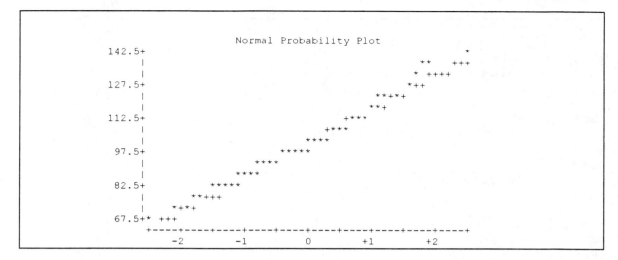

The third section, labeled Extreme Cases, lists the five lowest and five highest observations in your dataset. Examining these points is a good starting place (and often a finishing place) in your examination of the data for outliers.

Next, we come to the plots we requested with the "plot" option. As you can see SAS UNIVARIATE provides both a stem-and-leaf display and a box-plot placed side-by-side, so that you can use them together to examine the shape of your distribution. They use the same vertical scale, so the list of frequencies running down the middle applies to both of them. In PROC UNIVARIATE, the box-plot is marked with a horizontal line at the median value, and a plus sign "+" at the mean. Values more than 1.5 times the interquartile range from the edges of the box are marked with a zero "0" while SAS indicates extreme outliers, values more than 3 times the interquartile range distant, with an asterisk "*". Fortunately, no such points are evident in this sample.

Finally, there is a plot labeled Normal Probability Plot. These plots were not covered in the textbook but their purpose is to provide another visual test for normality of your data. The row of plus signs "+++++" rising on a diagonal from lower left to upper right represent a normal curve generated using your data (never mind how, for now). The asterisks "****" represent your actual data. Where the two symbols overlap, only the asterisks will be visible. As long as the asterisks don't deviate too far from the plus signs, you can be reasonably sure that your data is approximately normal.

In conclusion, PROC UNIVARIATE is a powerful procedure which will usually be your best starting place for examining your data.

Descriptives by Groups

Occasionally you have groups in your data and wish to examine them separately. In this case you need only sort the data set by the variable with the group information, and then use a By statement or Class statement in any other procedures to obtain separate analyses for each group. The following example is taken from **DS02_04** and corresponds to the data for six- and eight-year-olds on a vocabulary test reported in Section 2.10 of the textbook.

We are using the FORMAT procedure and statement for their more usual use, here, supplying informative names, usually called Value Labels, for categorical data that is coded as meaningless numbers. We could apply a similar methodology for a data set which has Gender coded as 0 for males and 1 for females, for example. After loading the data, we are sorting it by our grouping variable. We then request some descriptive statistics using the Univariate procedure and a pair of side-by-side boxplots using the By statement with this procedure. The results of this procedure are given on pages 175–179.

```
Options ls=80 ps=60;

Proc Format;
        Value Groupfmt 1='Six-Year-Olds'
                       2='Eight-Year-Olds';

Data Vocab;
        Input group correct;

Lines;
>>>Data Omitted<<<
Run;

Proc Sort;
        By Group;
Run;

Proc Univariate plot;
        Format Group Groupfmt.;
        By Group;
        Var Correct;
Run;
```

Note that we have obtained separate descriptive statistics and plots for both groups in our data set, and then finally a pair of full page box-plots so that we can directly compare the two groups. In order to obtain a large full page box-plot when we have no groups in our data it is necessary to invent a grouping variable in the data step and then assign it the same value for all subjects. When we perform our PROC UNIVARIATE By this artificial grouping variable we will obtain all the usual analyses plus the full page box-plot for all the data.

```
The SAS System                              5
                                              09:20 Thursday, March 27, 199

       ---------------------------- GROUP=Six-Year-Olds ---------------------------

                           Univariate Procedure

Variable=CORRECT


                                 Moments

                N              50    Sum Wgts            50
                Mean        0.601    Sum              30.05
                Std Dev  0.167554    Variance      0.028074
                Skewness 0.112773    Kurtosis      -0.29908
                USS       19.4357    CSS            1.37565
                CV       27.87927    Std Mean      0.023696
                T:Mean=0 25.36317    Pr>|T|          0.0001
                Num ^= 0       50    Num > 0             50
                M(Sign)        25    Pr>=|M|         0.0001
                Sgn Rank    637.5    Pr>=|S|         0.0001


                           Quantiles(Def=5)

               100% Max     0.95        99%       0.95
                75% Q3      0.69        95%       0.89
                50% Med    0.585        90%      0.855
                25% Q1      0.51        10%       0.35
                 0% Min     0.27         5%       0.33
                                         1%       0.27

               Range        0.68
               Q3-Q1        0.18
               Mode         0.55


                                Extremes

               Lowest    Obs      Highest    Obs
                0.27(     43)      0.88(      27)
                0.28(     14)      0.88(      50)
                0.33(     39)      0.89(      20)
                0.34(     30)      0.93(       8)
                0.34(      3)      0.95(       9)
```

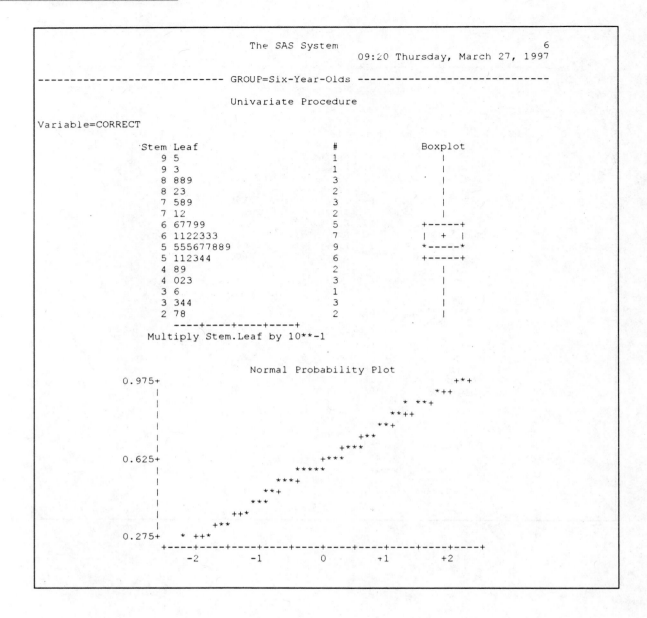

```
                          The SAS System                        7
                                      09:20 Thursday, March 27, 1997

-------------------------- GROUP=Eight-Year-Olds --------------------------

                        Univariate Procedure

Variable=CORRECT


                            Moments

            N              50    Sum Wgts           50
            Mean        0.7672   Sum             38.36
            Std Dev   0.145785   Variance     0.021253
            Skewness  -0.72825   Kurtosis     0.016077
            USS        30.4712   CSS          1.041408
            CV        19.0022    Std Mean     0.020617
            T:Mean=0  37.21184   Pr>|T|         0.0001
            Num ^= 0        50   Num > 0           50
            M(Sign)         25   Pr>=|M|        0.0001
            Sgn Rank     637.5   Pr>=|S|        0.0001


                        Quantiles(Def=5)

            100% Max     0.98      99%      0.98
             75% Q3      0.87      95%      0.96
             50% Med     0.8       90%     0.945
             25% Q1      0.67      10%      0.55
              0% Min     0.42       5%      0.45
                                    1%      0.42
            Range        0.56
            Q3-Q1        0.2
            Mode         0.8


                          Extremes

            Lowest     Obs     Highest    Obs
            0.42(       48)     0.95(      28)
            0.43(        6)     0.95(      40)
            0.45(       41)     0.96(       8)
            0.48(       10)     0.98(      17)
            0.53(       23)     0.98(      50)
```

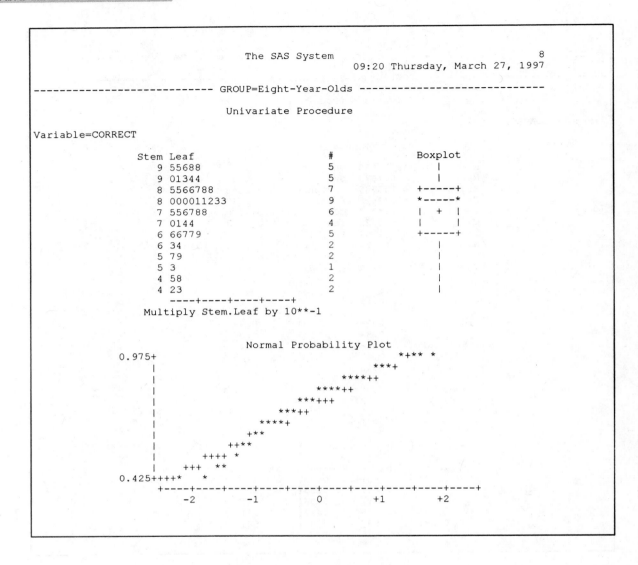

```
                              The SAS System                           8
                                    09:20 Thursday, March 27, 1997

--------------------------- GROUP=Eight-Year-Olds ----------------------------

                          Univariate Procedure

Variable=CORRECT

            Stem Leaf                      #          Boxplot
               9 55688                      5             |
               9 01344                      5             |
               8 5566788                    7          +-----+
               8 000011233                  9          *-----*
               7 556788                     6          |  +  |
               7 0144                       4          |     |
               6 66779                      5          +-----+
               6 34                         2             |
               5 79                         2             |
               5 3                          1             |
               4 58                         2             |
               4 23                         2             |
                 ----+----+----+----+
            Multiply Stem.Leaf by 10**-1


                        Normal Probability Plot
        0.975+                                      *+** *
             |                                   ***+
             |                               ****+
             |                             ****++
             |                          ***+++
             |                       ***++
             |                    *****+
             |                  +**
             |               ++**
             |            +++*  *
             |         ++++  *
             |       +++   **
        0.425++++*    *
             +----+----+----+----+----+----+----+----+----+----+
                 -2        -1        0        +1        +2
```

```
                              The SAS System                        9
                                      09:20 Thursday, March 27, 1997

                           Univariate Procedure
                             Schematic Plots

Variable=CORRECT

                            |
                  1 +       |
                            |                        |
                            |       |                |
                            |       |                |
                0.9 +       |       |                |
                            |       |            +-----+
                            |       |            |     |
                            |       |            |     |
                0.8 +       |       |            *-----*
                            |       |            |  +  |
                            |       |            |     |
                            |       |            |     |
                0.7 +       |    +-----+         |     |
                            |    |     |         +-----+
                            |    |     |            |
                            |    |     |            |
                0.6 +       |    |  +  |            |
                            |    *-----*            |
                            |    |     |            |
                            |    |     |            |
                0.5 +       |    +-----+            |
                            |       |               |
                            |       |               |
                            |       |               |
                0.4 +       |       |               |
                            |       |
                            |       |
                0.3 +       |       |
                            |       |
                            |
                0.2 +       ------------+----------+----------
                  GROUP        Six-Year    Eight-Year
```

Before we leave descriptive statistics, it is worth noting that there is another SAS procedure that produces numerical descriptions of our data sets. The procedure is called MEANS and it is just as easy to use as PROC UNIVARIATE though its output is somewhat easier to read. Note that you cannot obtain any plots from PROC MEANS. The syntax is as follows:

```
Proc Means;
      Var Correct;
   Run;
```

This produces the following default output when applied to our six- and eight-year-olds:

```
The SAS System                          10
                                  09:20 Thursday, March 27, 1997

        Analysis Variable : CORRECT

      N        Mean       Std Dev      Minimum      Maximum
    ------------------------------------------------------------
     100     0.6841000    0.1771725    0.2700000    0.9800000
```

Note that if we has wished separate analyses for our two groups we could have used a By statement just as we did with PROC UNIVARIATE. It is also possible to request different summary statistics from PROC MEANS by specifying the requested statistic as a procedure option. If you do so, you will not obtain the default output above, but only the statistics you request. Some available statistics you might find useful are: N, NMISS (number of missing values), MEAN, STD (standard deviation), STDERR (standard error on the mean), MIN, MAX, VAR (variance), SKEWNESS, and KURTOSIS. An example of code requesting some of these options follows:

```
Proc Means N NMISS STDERR SKEWNESS;
      Var Correct;
    Run;
```

Review of Concepts

■ SAS can be used to obtain many numerical and graphical descriptions of your data.

■ Frequency statistics are available from PROC FREQ. You can obtain tables of frequencies using the Tables statement. Bar charts and Histograms of your data are available from PROC CHART, though PROC GCHART produces nicer, high-resolution graphs. In both cases you use the Vbar statement to define your chart.

■ You can vary the number of class intervals in your bar charts and histograms by using the Levels or Midpoints options on the Vbar statement. Additionally, you can create new class variables in the Data statement. Sometimes recoded data will not be plotted as you'd expected, you may need some trial and error experimentation to get things looking the way you'd hoped.

■ PROC UNIVARIATE is a powerful tool for obtaining both numerical and graphical descriptions of your data. In Univariate you can select many different statistics and graphs for presentation, but even the default options will usually be useful. PROC MEANS is another way to obtain many of the same numerical descriptions.

■ By sorting your data by a group variable in PROC SORT, you gain the option of using it in a By statement in PROC UNIVARIATE and PROC MEANS. By doing so you can obtain separate numerical and graphical descriptions for each group in your data. You can also perform side-by-side comparisons of the groups with the box-plots produced. Always remember to provide informative Value Labels for your group variable(s) by using PROC FORMAT and the Format statement.

Exercises

1. Enter Data Set 2.5 from the keyboard or from the file **DS02_05.dat**, noting that this is a "string" variable. Remember that suggested variable names can be found in the listing of data sets in Appendix B of this workbook. Obtain a frequency distribution of the data.

2. Look at Problem 2 in Section 2.2 of the textbook. The data for this problem can be found in the file **DS02_06.dat**. Obtain descriptive statistics for the sample as a whole and separately for each group. How do the groups compare to one another numerically and graphically?

3. Using data in the file **DS02_07.dat** from Problem 3 of Section 2.1, obtain a stemplot, the mean, median, and standard deviation of the data.

4. Using data in the file **DS02_08.dat**, check your answers to Problem 5 of Section 2.2.

5. Using data in the file **DS02_09.dat**, obtain the variance of each data set used in Problem 8 of Section 2.2.

6. Using data in the file **DS02_10.dat**, obtain the standard deviations of each data set used in Problem 9 of Section 2.2.

3 OBTAINING STANDARD ERRORS AND CONFIDENCE INTERVALS

In Section 2 of this workbook we left off our discussion of descriptive statistics having obtained separate descriptions for two different groups in our data set, as well as a graphical comparison of the groups using side-by-side box-plots. Section 3 explores methods in SAS for determining whether the means of two groups really differ. Students who feel the need to review the notion of group differences should consult Chapters 5 and 6 of the textbook.

The example we will be working with is described in Section 6.2 of the text. It consists of data from an experiment on the effectiveness of desensitization therapy for the reduction of anxiety in patients with phobias. The data can be found in file **DS03_01.dat**. The first column consists of the group information, with a zero indicating that that subject was part of the control group in this study, and a one indicating that they received the desensitization therapy. The second column is a measure of anxiety, the recorded increase in the patient's heart rate upon presentation of the phobic stimulus, after completion of the therapy (or control). We used the following Data Step to load the data into SAS as described in Section 1, calling the first variable Group and the second Anxiety.

In order to facilitate the interpretation of this data set, we provide some informative labels for the variables. The Label statement is used in the Data step to produce Variable Labels while PROC FORMAT and the Format statement are used to supply Value Labels. For the Group variable we supplied the Variable Label "Therapy" and the Value Labels "control" if group = 1 and "desensitization" if group =2. For the Anxiety variable we added the Variable Label "Increase in Heart Rate", but we did not add value labels as Anxiety was a continuous variable. All of this should be familiar from Section 2 of the SAS module of this workbook, with the exception of Variable Labels. Variable Labels are a useful way to provide additional information about the nature of a variable which for the sake of convenience you will load under a short name. You will find Variable Labels to be particularly useful when you return to your data after a long interval and cannot remember which measure of anxiety you had used in a particular study (see page 183).

Having processed the data, we have invoked PROC UNIVARIATE, using Anxiety as the variable of interest. Prior to running PROC UNIVARIATE we sorted the data by GROUP so that we could use Group in the By statement for the Univariate procedure. This allows us to compare numerical and graphical descriptors of Anxiety between the two experimental groups. The results of this program are presented on pages 184–186. Only the Plot option of PROC UNIVARIATE was used.

```
Options ls=80 ps=60;

Proc Format;
        Value Exptfmt 1='control' 2='desensitization';
Run;

Data Therapy;
        Input Group Anxiety;

        Label Group='Therapy'
             Anxiety='Increase in Heart Rate';

        Format Group Exptfmt.;

Lines;
>>>Data Omitted<<<
Run;

Proc Sort;
        By Group;
Run;

Proc Univariate plot;
        Var Anxiety;
        By Group;
Run;
```

Examining the descriptive statistics, we note that the mean increase in heart rate for the control group in the stressful situation was greater than for the desensitization group. We also note that the latter group has a greater spread reflected in its higher variance. Both these observations are visible in both the stem-and-leaf plots and the box-plots. The question is whether we have any principled reason to believe that the means would really be different if we collected another sample, or simply enlarged this one. In Chapters 5 and 6 of the text book you are presented with a method of addressing this question based on confidence intervals. While SAS does not directly output 95% (or whatever) confidence intervals on the means of each group, for immediate comparison, you can see that PROC UNIVARIATE provides all the necessary data for you to calculate them using the means and standard errors that are output. We leave it as an exercise for the interested reader to calculate these confidence intervals to determine whether the two groups really have different means, see Chapter 5.2 in the textbook for assistance. We turn our attention to statistical approaches using hypothesis testing to answer this same question.

In Chapter 6 of the text we learned that an appropriate statistical method to test the difference between the means of two groups on a given measure was the *t*-test. SAS output uses uppercase T, and so the procedure is referred to as the T-Test. T-Tests are just one way of examining such a difference, but the one which we will limit ourselves to here. As our group variable represents two independent groups of patients, we wish to use an Independent Samples T-Test.

To obtain such a test in SAS we invoke PROC TTEST. There are two required statements in PROC TTEST, a Class statement and a Var statement. The Var statement should be familiar to you from other procedures we invoked above, and in Section 2. It is simply the place to list the response variable(s) of interest in your study, in this case Anxiety. If more than one variable is specified in the Var statement, separate T-Tests will be performed on each, using the Class variable as the predictor (independent) variable in each case. The Class statement is where you list the grouping or predictor variable

```
The SAS System                              1
                                02:12 Tuesday, April 8, 1997

----------------------------- Therapy=control ----------------------------------

                        Univariate Procedure

Variable=ANXIETY       Increase in Heart Rate

                              Moments

               N              18   Sum Wgts          18
               Mean         10.2   Sum             183.6
               Std Dev  1.873028   Variance     3.508235
               Skewness -0.42328   Kurtosis     1.671839
               USS       1932.36   CSS             59.64
               CV       18.36302   Std Mean     0.441477
               T:Mean=0 23.10426   Pr>|T|         0.0001
               Num ^= 0       18   Num > 0           18
               M(Sign)         9   Pr>=|M|        0.0001
               Sgn Rank     85.5   Pr>=|S|        0.0001


                           Quantiles(Def=5)

           100% Max        14      99%           14
            75% Q3       11.3      95%           14
            50% Med      10.3      90%         12.6
            25% Q1        9.4      10%          8.2
             0% Min       5.5       5%          5.5
                                    1%          5.5

           Range          8.5
           Q3-Q1          1.9
           Mode          10.8


                              Extremes

           Lowest     Obs      Highest     Obs
             5.5(       5)       11.3(       9)
             8.2(       3)       11.3(      14)
             8.4(       1)       12.1(      18)
             8.9(      10)       12.6(       6)
             9.4(      16)         14(       8)


      Stem Leaf                      #        Boxplot
        14 0                         1           |
        13                                       |
        12 16                        2           |
        11 33                        2        +-----+
        10 15688                     5        *--+--*
         9 4678                      4        +-----+
         8 249                       3           |
         7
         6
         5 5                         1           0
           ----+----+----+----+
```

```
                              The SAS System                                    2
                                            02:12 Tuesday, April 8, 1997

---------------------------- Therapy=control --------------------------------

                           Univariate Procedure

Variable=ANXIETY        Increase in Heart Rate

                         Normal Probability Plot
           14.5+                                        *   +++++
               |                                      ++++++
               |                               +*+*+
           11.5+                           ++*+
               |                     *+**+*
               |                  *+**+*
            8.5+            *  *++*+
               |         ++++++
               |     +++++
            5.5+++++     *
               +----+----+----+----+----+----+----+----+----+----+
                  -2        -1         0        +1        +2
```

```
                              The SAS System                                    3
                                            02:12 Tuesday, April 8, 1997

-------------------------- Therapy=desensitization --------------------------

                           Univariate Procedure

Variable=ANXIETY        Increase in Heart Rate

                                Moments

                 N              18    Sum Wgts              18
                 Mean          5.2    Sum                 93.6
                 Std Dev  2.660164    Variance        7.076471
                 Skewness -0.37737    Kurtosis        -0.96594
                 USS        607.02    CSS                120.3
                 CV       51.15699    Std Mean        0.627007
                 T:Mean=0 8.293374    Pr>|T|            0.0001
                 Num ^= 0       17    Num > 0             17
                 M(Sign)       8.5    Pr>=|M|           0.0001
                 Sgn Rank     76.5    Pr>=|S|           0.0001


                           Quantiles(Def=5)

                 100% Max       8.8        99%         8.8
                  75% Q3        7.8        95%         8.8
                  50% Med      5.35        90%         8.5
                  25% Q1        2.4        10%         1.9
                   0% Min         0         5%           0
                                            1%           0

                 Range          8.8
                 Q3-Q1          5.4
                 Mode           1.9


                                Extremes

                 Lowest    Obs      Highest    Obs
                      0(    14)        7.8(    15)
                    1.9(    10)        7.9(     9)
                    1.9(     3)        8.1(    18)
                    2.3(     7)        8.5(     2)
                    2.4(    16)        8.8(     5)
```

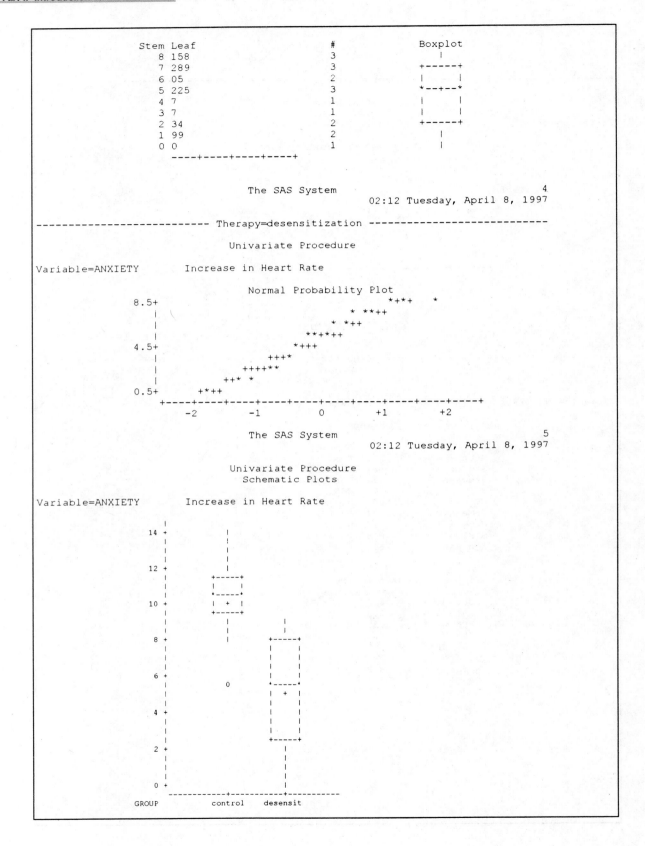

for the T-Test, in our example Group. The Class variable must contain only two values, like the 1 and 2 values in our Group variable, for PROC TTEST to work. Note that unlike the By statement in PROC UNIVARIATE above, the variable named in a Class statement need not be sorted in advance. Nonetheless, you will usually want to examine some descriptive statistics for each group in your data anyway, so sorting by your predictor variable is virtually always needed.

```
Proc Ttest;
      Class Group;
      Var Anxiety;
    Run;
```

The results of this procedure follow.

```
TTEST PROCEDURE

Variable: ANXIETY        Increase in Heart Rate

             GROUP      N          Mean            Std Dev         Std Error
------------------------------------------------------------------------------
           control      18      10.20000000       1.87302838       0.44147702
desensitization         18       5.20000000       2.66016364       0.62700658

Variances       T        DF      Prob>|T|
----------------------------------------------
Unequal      6.5203     30.5      0.0001
Equal        6.5203     34.0      0.0000

         For H0: Variances are equal, F' = 2.02    DF = (17,17)    Prob>F' = 0.1581
```

Note that the output begins with a label for the variable under analysis. It is always a good idea to double check this, as we advised in Section 2. Note that once you begin conducting multiple analyses on the same or similar data sets, the helpful Variable Labels we provided with the Labels statement, will become invaluable, as you struggle to keep your many pages of output organized.

We are next presented with some descriptive statistics for each of our groups. Double check the Ns to make sure that SAS and you agree on the number of subjects in your groups. It is also worthwhile pausing to note which group has the larger mean, and which the larger variance at this point. Following these, there are a pair of T-Tests labeled Unequal and Equal respectively. You should consult the line labeled Unequal if you have reason to believe that your two groups have different variances, and Equal otherwise. For both tests you are provided with a T value, and a probability of attaining a T value this extreme or greater under the null hypothesis. If the number under Prob>|T| is less than 0.05 you have reason to not retain the null hypothesis as an option in deciding whether your means are equal. In other words, if you reject the null hypothesis, you are claiming that your means are not equal, and can thus state that you have reason to believe that the condition with the larger mean, truly has a larger mean than the condition with the smaller mean. In this case we conclude that the group receiving the desensitization therapy, responded to the phobic situation with less increase in heart rate, and thus presumable less anxiety. We have marshaled one piece of evidence in favor of belief that our therapy is successful. If you need some refreshment on the logic of hypothesis testing you should consult Chapter 6.4 in the textbook.

The final lines of the output present you with a statistical test of whether your groups have equal variances. The test is for Equality of Variances and is usually represented by F' as in this case. It is a test of the null hypothesis that the variances are equal, and thus if it is not significant at the 0.05 level we have no reason to suspect heterogeneity of variance problems, and should use the T-Test labeled Equal Variances above. In our example the obtained probability was given as Prob>F′ = 0.1581, and so we can use the Equal Variances T value.

Review of Concepts

■ You can compare differences between two groups in your data using descriptive statistics and confidence intervals or independent samples T-tests.

■ When importing your data it is a good idea to use Variable Labels, using the Labels Statement in the Data Step, to give your variables longer, more descriptive names than you can use while processing them, for future reference. You should also assign Value Labels to the levels of your grouping variable, using PROC FORMAT and the Format Statement.

■ Statistics obtained from PROC UNIVARIATE can be used to compare your two groups if you previously sort the data set by the grouping variable using PROC SORT, and then add a By statement to PROC UNIVARIATE. This procedure was first introduced in Section 2.

■ You can use the output of PROC UNIVARIATE to calculate confidence intervals on means of your groups, and to examine differences in the groups means and variances numerically and graphically.

■ PROC TTEST allows you to perform an independent samples T-test on the data, as well as providing useful descriptives and a test for homogeneity of variance among your groups.

Exercises

1. Import the tranquilizer data set described in Section 6.22 of your textbook. It can be found in ASCII file **DS06_01.dat**. Assign Variable Labels and Label Statements. Use PROC UNIVARIATE to examine the data from each condition, then use PROC TTEST to obtain a 95% confidence interval for the difference between the means of the conditions and perform an independent *t*-test. What conclusion about the effects of the tranquilizer would you draw from this study?

2. Repeat the steps described in Exercise 1 for the following data sets.

   a. The first worked example at the end of Section 6.2 (induced happiness experiment). The data are found in Data Set 6.2 (ASCII file **DS06_02.dat**).

   b. The first Problem at the end of Section 6.2 (induced anger experiment). The data are found in Data Set 6.3 (ASCII file **DS06_03.dat**).

   c. Problem 2 at the end of Section 6.2 (intentional versus incidental remembering experiment). The data are found in Data Set 6.4 (ASCII file **DS06_04.dat**).

4 ANALYSIS OF VARIANCE FOR INDEPENDENT GROUPS

Chapter 7 of the textbook describes the analysis of variance (abbreviated ANOVA) as a procedure for analyzing the results of experiments with more than two levels of a categorical predictor variable. In Section 7.1 you were introduced to the idea of estimating differences between the means of the levels of the predictor variable using both point estimates and confidence intervals, based on the distribution of a t statistic and on the distribution of the Studentized range statistic, q. In Section 7.2 you were introduced to a method for simultaneously evaluating differences among the group means by generating a sum of squares for the model and for the null hypothesis, dividing these sums of squares by their degrees of freedom, and taking a ratio of the resulting variances. We begin our discussion experiments with more than two levels using this latter technique, the analysis of variance, and then proceed to look at ways to examine differences between group means directly.

The basic procedure for the analysis of variance in SAS is PROC ANOVA. It has two required statements. A CLASS statement, in which the categorical response variable is listed, and a MODEL statement, in which the model relating the predictor variable to the response variable is stated. Below is an example of a basic program to perform a simple ANOVA using the data presented in Section 7.1 of the textbook. You can find these data in the ASCII file **DS07_01.dat**. The data report the heart rate of phobic patients presented with their phobic object, after receiving one of three approaches to treatment. For the purposes of this example we will call the predictor variable Group and the response variable Heartrt.

```
DATA Section4;

        INFILE 'C:\datasets\DS07_01.dat';
        INPUT Group Heartrt;
RUN;

PROC ANOVA data=Section4;
        CLASS Group;
        MODEL Heartrt=Group;
RUN;
```

As you can see the ANOVA procedure is fairly simple to implement, as long as you understand the underlying model being used. In this case the MODEL statement only includes two terms, one for the response variable Heartrt, which is said to be equal to the predictor variable GROUP. This is equivalent to telling SAS that we believe our data has a structure equivalent to: $Y_{ij} = \mu + \alpha_i + e$. SAS takes it for granted that there is an effect of the grand mean, though it won't actually test for one unless we specify a option in the model statement. Likewise, it is assumed that there is error variance within each group, unless a within-subjects (repeated measure) analysis is being conducted (see Chapter 8 of the textbook, Section 5 of this workbook for more details.). As a result, when SAS presents an ANOVA summary table for this model, the component being tested for is the effect of Group. The results from running this procedure follow.

```
The SAS System                                                          1

                    Analysis of Variance Procedure
                    Class Level Information

                 Class     Levels     Values

                 GROUP        3        1 2 3

       Number of observations in data set = 45
                                                                        2

                         The SAS System


                    Analysis of Variance Procedure

Dependent Variable: HEARTRT
                              Sum of          Mean
Source              DF       Squares         Square    F Value    Pr > F

Model                2   210.00000000   105.00000000    16.33     0.0001

Error               42   270.00000000     6.42857143

Corrected Total     44   480.00000000

           R-Square          C.V.        Root MSE        HEARTRT Mean

           0.437500        36.22090      2.5354628         7.0000000

Source              DF       Anova SS    Mean Square    F Value    Pr > F

GROUP                2   210.00000000   105.00000000    16.33     0.0001
```

The first page of the output provides summary information about the variables being used in the ANOVA. It is always a good idea to double check this information to make sure that you and SAS agree about how many levels your predictor variable has, the size of your sample, etc. The second page of output presents the results of the analysis in the form a an ANOVA summary table. The description begins with the name of the response variable, and their follows a table which displays the sum of squares for your model, as well as the residual, plus their associated degrees of freedom, mean squares and the F-ratio of these, as well as the probability value associated with this magnitude of F at these degrees of freedom. For our example above, we would report $F(2,42) = 16.33$, $p<0.0001$. In other words there is at most a 1 in $10,000^{th}$ chance that data this extreme were produced by a true H_0. We conclude therefore that we have sufficient evidence to discredit the null hypothesis that the means of our three groups are equal.

The next line of output presents several statistics which are primarily of interest from the perspective of regression analysis, of which ANOVA can be considered a specialized example. For now we will only note the value for R-square (R^2), which for our example is 0.4375, is a measure of effect size obtained by dividing the sum of squares for the model by the total sum of squares. This can be considered a measure of association, or alternately a measure of the proportion of variance explained by our model. In other words, we can account for approximately 44 percent of the variability in any subject's Heartrt by knowing to which Group they belong. In the context of ANOVA this statistic is commonly called η^2 (the Greek letter is eta, pronounced ee-ta). This statistic has fallen into some disfavour lately, as it is positively biased in many cases, however is still the most common measure of effect size reported in the literature for ANOVA results.

The next line of the output contains the model line from the above table, only this time listed by the variable name, Group. In our example, the two lines are identical in all respects, and you should feel free to examine either. If we had performed a factorial ANOVA, an ANOVA with more than 1 predictor variable, however; the two sections of the table would differ. The upper portion of the table would present the overall results for the model, with all predictor variables taken into account, while the lower portion of the table, would present the individual analyses for each predictor variable. Thus the sum of squares and degrees of freedom for the second half of the table would sum to the sum of squares and degrees of freedom for "model" in the first half. Of course it does even in our example, but with only 1 variable this is trivially true.

The significant results obtained for Group, in the above output only tells us that the means of the three groups are not the same. It tells us nothing about which means are larger, or indeed which means differ from one another. To further examine questions of differences among the treatment means, it is necessary to engage in a comparison procedure which will control the error rate per experiment. The textbook identified two approaches to this problem. One could test the differences among the means independently of any ANOVA using either a Bonferroni procedure to control the Type-1 error rate of each t-test, or Tukey's use of the Studentized range statistic, q, to perform simultaneous comparisons. The second option, was to run the ANOVA, and then if one obtained significant results, to examine comparisons among pairs of means, using t-tests. This second procedure was known as Fisher's LSD. Unfortunately this distinction is somewhat blurred in SAS by the fact that any of these comparisons can only be requested after an ANOVA has been run. To request a comparison of group means one merely adds the optional statement Means to an ANOVA procedure. This statement must follow the MODEL and CLASS statements. An example follows:

```
MEANS Group / Tukey LSD REGWQ;
```

The syntax is to use the statement Means, followed by the name of the effect we wish to examine in greater detail (in our example, the effect of Group). If we left the statement at this point, SAS would merely print out the means of each level of Group. Adding the options after the slash, requests additional procedures be carried out, to test the comparisons among pairs of means. In this case, we have selected Tukey (the SAS command for Tukey's HSD) and LSD to allow for comparison between these two procedures. Normally you would only select one, however. In addition to these two options, one could select others, including the Bonferroni test (using the command BON) and simple t-tests (using the command T). In addition to the two tests we have discussed, we have included a command for a third comparison procedure called REGWQ in SAS for the first initials of the four statisticians who have contributed to its development: Ryan, Einot, Gabriel and Welch. From here on we will refer to it as the

Ryan procedure, after the person who first suggested this approach. Although not covered in the textbook, it is mentioned here for those who may be interested. The Ryan procedure is similar to Tukey's HSD in that it tests means based on a value of the Studentized range statistic, q. Unlike Tukey, however, it uses a different range for each comparison and in this way is like the older, now discredited, Newman-Keuls procedure. Unlike that earlier procedure, however, Ryan adjusts the alpha level of each comparison to take into account the number of comparisons being made. It thus offers an excellent compromise between the Tukey procedure (which most feel is too conservative) and Fisher's LSD or Newman-Keuls (which most feel are too lenient). It is likely to become more popular in coming years, though it is too computationally cumbersome to carry out by hand. As of this writing SAS and SPSS are the only statistical packages to support this innovative technique, and thus we present it for your consideration, along with the more traditional techniques. The results from adding this MEANS statement to the above analysis are shown on page 193.

As can be seen the output for each of these post-hoc comparisons has a similar format. In each case, following the name of the procedure employed you are presented with a brief description of the control on error rate employed by each of the tests. Fisher's LSD is reported as controlling the error rate per comparison, rather than per experiment. This indicates that with large numbers of comparisons (i.e. A study with many levels of the grouping variable), the alpha rate for the experiment may become unacceptably high. Tukey's HSD controls the error rate per experiment by furnishing a minimum difference between means based on the largest range of ordered means possible in the experiment. While this controls Type-1 errors, it is too conservative and thus we are warned of the increased chance of making a Type-2 error. The Ryan procedure controls error rate per experiment by furnishing a minimum difference between means based on the range of each pair of ordered means, using a Bonferroni-like procedure to adjust the alpha level for each comparison. Thus control over the possibility of Type-1 errors is maintained while not increasing the likelihood of a Type-2 error.

After these descriptions you are presented with the critical statistics associated with each test including the critical differences between the means based upon each procedure, and then a listing of the means of each level of Group, ordered from largest to smallest. Next to the list of means is a column of letters. For each procedure, means with the same letter next to them, do not differ, while means with different letters next to them do. In our example, all three procedures agree in their conclusions. The mean of Group 1 is significantly greater than the means of Group 2 and Group 3. Group 2 and Group 3 do not significantly differ from each other.

Chapter 7 began with an examination of differences among groups of means by constructing 95% confidence intervals around these differences, based upon either the Studentized range statistic, q. or the t statistic. In order to have SAS produce these confidence intervals for you, you simply preface your optional comparison procedures with the statement Cldiff. For example:

```
MEANS Group / Cldiff Tukey LSD;
```

This produces the output on pages 194–195, in which the differences between pairs of means are reported as confidence intervals, based first upon the Studentized range statistic (requested by the Tukey option above) and then upon t (requested by LSD).

Note that once you have toggled the Means statement to present differences as confidence intervals, it no longer presents the default output of the testing procedures, as seen above. Thus in order to obtain both confidence intervals and significance tests

```
The SAS System                          3

                  Analysis of Variance Procedure

                 T tests (LSD) for variable: HEARTRT

   NOTE: This test controls the type I comparisonwise error rate not the
         experimentwise error rate.

              Alpha= 0.05  df= 42  MSE= 6.428571
                   Critical Value of T= 2.02
              Least Significant Difference= 1.8684

   Means with the same letter are not significantly different.

              T Grouping          Mean      N  GROUP

                     A       ·   10.0000    15  1

                     B           6.0000     15  2
                     B
                     B           5.0000     15  3


                      The SAS System                          4

                  Analysis of Variance Procedure

   Ryan-Einot-Gabriel-Welsch Multiple Range Test for variable: HEARTRT

      NOTE: This test controls the type I experimentwise error rate.

              Alpha= 0.05  df= 42  MSE= 6.428571

           Number of Means          2         3
           Critical Range   1.8684065 2.2492752

   Means with the same letter are not significantly different.

            REGWQ Grouping          Mean      N  GROUP

                     A           10.0000    15  1

                     B           6.0000     15  2
                     B
                     B           5.0000     15  3


                      The SAS System                          5

                  Analysis of Variance Procedure

     Tukey's Studentized Range (HSD) Test for variable: HEARTRT

   NOTE: This test controls the type I experimentwise error rate, but
         generally has a higher type II error rate than REGWQ.

              Alpha= 0.05  df= 42  MSE= 6.428571
           Critical Value of Studentized Range= 3.436
            Minimum Significant Difference= 2.2493

   Means with the same letter are not significantly different.

            Tukey Grouping          Mean      N  GROUP

                     A           10.0000    15  1

                     B           6.0000     15  2
                     B
                     B           5.0000     15  3
```

```
                            The SAS System                              3

                      Analysis of Variance Procedure


T tests (LSD) for variable: HEARTRT

        NOTE: This test controls the type I comparisonwise error rate not the
              experimentwise error rate.

        Alpha= 0.05  Confidence= 0.95  df= 42  MSE= 6.428571
                      Critical Value of T= 2.01808
                Least Significant Difference= 1.8684

Comparisons significant at the 0.05 level are indicated by '***'.

                          Lower     Difference    Upper
              GROUP     Confidence    Between    Confidence
            Comparison    Limit       Means        Limit

             1   - 2      2.1316      4.0000       5.8684      ***
             1   - 3      3.1316      5.0000       6.8684      ***

             2   - 1     -5.8684     -4.0000      -2.1316      ***
             2   - 3     -0.8684      1.0000       2.8684

             3   - 1     -6.8684     -5.0000      -3.1316      ***
             3   - 2     -2.8684     -1.0000       0.8684
```

```
                            The SAS System                              4

                      Analysis of Variance Procedure

        Tukey's Studentized Range (HSD) Test for variable: HEARTRT

        NOTE: This test controls the type I experimentwise error rate.

        Alpha= 0.05  Confidence= 0.95  df= 42  MSE= 6.428571
              Critical Value of Studentized Range= 3.436
                Minimum Significant Difference= 2.2493

Comparisons significant at the 0.05 level are indicated by '***'.

                       Simultaneous             Simultaneous
                          Lower     Difference     Upper
              GROUP     Confidence    Between    Confidence
            Comparison    Limit       Means        Limit

             1   - 2      1.7507      4.0000       6.2493      ***
             1   - 3      2.7507      5.0000       7.2493      ***

             2   - 1     -6.2493     -4.0000      -1.7507      ***
             2   - 3     -1.2493      1.0000       3.2493

             3   - 1     -7.2493     -5.0000      -2.7507      ***
             3   - 2     -3.2493     -1.0000       1.2493
```

of the post-hoc comparisons, it would be necessary to have more than one MEANS statement in your ANOVA procedure.

Factorial ANOVA

In Section 7.3 of the textbook you were introduced to a method for testing differences among means in experiments involving two (or more) predictor variables using Factorial ANOVA. In a two way ANOVA we assume that each data point can be modeled by a linear combination of a main effect for each of the two predictor variables plus an interaction term for the effects of the unique combinations of these predictor variables, plus , of course, a residual component. An example of such an experiment is given in Section 7.3, involving the effect on driving performance of two different levels of sleep deprivation, in combination with two different dosages of alcohol. That data for this experiment can be found in the data file **DS07_07.dat** in the *ASCII files*. A SAS program to analyze this data follows:

```
OPTIONS ls=80 ps=54;

DATA Factors;

        INFILE 'C:\datasets\DS07_07.dat';
        INPUT Alcohol Sleep Score;
RUN;

PROC ANOVA data=Factors;
        CLASS Alcohol Sleep;
        MODEL Score=Alcohol Sleep Alcohol*Sleep;
RUN;
```

As can be seen the sample program is virtually identical to the one we wrote for a one-way ANOVA above. The only differences are that we now have to list both predictor variables under the CLASS statement (if this seems ungrammatical to you, SAS will happily accept CLASSES), and we must write our model statement to reflect our new model of the data, listing both main effects and the interaction term. When we wish to perform a fully factorial ANOVA (the only kinds we have discussed in the textbook) SAS offers us a shorthand method of writing the model statement. Instead of separately listing Alcohol, Sleep and Alcohol*Sleep, in our model statement SAS would have treated as equivalent the statement,

```
MODEL Score=Alcohol|Sleep;
```

Where the vertical line | is taken to mean "all main effects and interactions for the terms joined by this line". The output from either variation of this ANOVA procedure is presented at the top of page 196.

Note that we again have our output divided into two sections, the first dealing with the model as a whole, and the second dealing with each treatment separately. In order to read this output, then, in a manner more familiar to you, you should look at the second section for the sums of squares, etc. of your treatments, and at the first section for your error term and total.

As can be seen both of our main effects are significant, but because we are dealing with a factorial design, we must first check to see if our interaction is significant. It is, and therefore we will have to interpret our main effects in light of this interaction. One way do this is to test for simple effects. Simple effects are the differences between means for one of our predictor variables, tested at each level of the other predictor variable. For example, we may wish to test the simple effect of Sleep Deprivation at each level of Alcohol Consumption. In other words we wish to perform two more one-way ANOVAs on this data, using half the data for each. To test the simple effect of Sleep Deprivation at each level of Alcohol Consumption we could write the program in the middle of page 196.

Note that we first have to sort our data set by the variable we wish to use in the BY statement. We have also added a MEANS statement to help us better understand the

```
The SAS System                              12:03 Tuesday, September 16, 1997

                        Analysis of Variance Procedure
                          Class Level Information

                        Class     Levels    Values

                        ALCOHOL      2       1 2

                        SLEEP        2       1 2


              Number of observations in data set = 40


                        Analysis of Variance Procedure

Dependent Variable: SCORE
                              Sum of          Mean
Source              DF        Squares         Square     F Value     Pr > F

Model                3       1.45400000     0.48466667    19.34      0.0001

Error               36       0.90200000     0.02505556

Corrected Total     39       2.35600000

              R-Square          C.V.        Root MSE         SCORE Mean

              0.617148        18.40575      0.1582895        0.8600000


Source              DF        Anova SS     Mean Square    F Value     Pr > F

ALCOHOL              1       0.72900000    0.72900000      29.10      0.0001
SLEEP                1       0.52900000    0.52900000      21.11      0.0001
ALCOHOL*SLEEP        1       0.19600000    0.19600000       7.82      0.0082
```

```
PROC SORT data=Factors;
      By Alcohol;
RUN;

PROC ANOVA data=Factors;
      CLASS Sleep;
      BY Alcohol;
      MODEL Score=Sleep;
      MEANS Sleep/ClDiff Tukey;
RUN;
```

mean differences of score for the Low and High Sleep Deprivation conditions under our two conditions of Alcohol dosage. The results of this program appear on pages 197–198.

As can be seen we have two complete outputs from our ANOVA procedure, one for the scores at each level of Alcohol consumption. By examining these simple effects analyses, the real pattern of results, giving rise to our initial interaction term is apparent. There is an effect of sleep deprivation (at the two levels employed) but only when subjects have had a drink.

```
-------------------------------- ALCOHOL=1 --------------------------------
                        Analysis of Variance Procedure
                          Class Level Information

                     Class     Levels    Values

                     SLEEP        2       1 2

            Number of observations in by group = 20



                          The SAS System                          12
                                12:03 Tuesday, September 16, 1997
-------------------------------- ALCOHOL=1 --------------------------------

                    Analysis of Variance Procedure

Dependent Variable: SCORE
                                Sum of          Mean
Source                DF        Squares         Square     F Value     Pr > F

Model                  1      0.04050000      0.04050000    1.26       0.2758

Error                 18      0.57700000      0.03205556

Corrected Total       19      0.61750000

              R-Square          C.V.        Root MSE          SCORE Mean

              0.065587        24.69526      0.1790407         0.7250000


Source                DF        Anova SS     Mean Square    F Value     Pr > F

SLEEP                  1      0.04050000     0.04050000      1.26       0.2758



                          The SAS System                          13
                                12:03 Tuesday, September 16, 1997
-------------------------------- ALCOHOL=1 --------------------------------

                    Analysis of Variance Procedure

        Tukey's Studentized Range (HSD) Test for variable: SCORE

      NOTE: This test controls the type I experimentwise error rate.

        Alpha= 0.05  Confidence= 0.95  df= 18  MSE= 0.032056
              Critical Value of Studentized Range= 2.971
              Minimum Significant Difference= 0.1682

      Comparisons significant at the 0.05 level are indicated by '***'.

                       Simultaneous              Simultaneous
                          Lower      Difference     Upper
              SLEEP     Confidence    Between    Confidence
           Comparison     Limit       Means        Limit

            2   - 1     -0.07822     0.09000      0.25822

            1   - 2     -0.25822    -0.09000      0.07822
```

```
                            The SAS System                          14
                              12:03 Tuesday, September 16, 1997

------------------------------- ALCOHOL=2 ------------------------------------

                        Analysis of Variance Procedure
                          Class Level Information

                    Class     Levels    Values

                    SLEEP        2       1 2

               Number of observations in by group = 20


                            The SAS System                          15
                              12:03 Tuesday, September 16, 1997

------------------------------- ALCOHOL=2 ------------------------------------

                        Analysis of Variance Procedure

Dependent Variable: SCORE
                                Sum of          Mean
Source                DF        Squares        Square     F Value     Pr > F

Model                 1       0.68450000     0.68450000    37.91      0.0001

Error                 18      0.32500000     0.01805556

Corrected Total       19      1.00950000

              R-Square          C.V.          Root MSE         SCORE Mean

              0.678058        13.50462        0.1343710         0.9950000


Source                DF        Anova SS      Mean Square   F Value     Pr > F

SLEEP                 1       0.68450000     0.68450000    37.91      0.0001


                            The SAS System                          16
                              12:03 Tuesday, September 16, 1997

------------------------------- ALCOHOL=2 ------------------------------------

                        Analysis of Variance Procedure

        Tukey's Studentized Range (HSD) Test for variable: SCORE

        NOTE: This test controls the type I experimentwise error rate.

             Alpha= 0.05  Confidence= 0.95  df= 18  MSE= 0.018056
                  Critical Value of Studentized Range= 2.971
                  Minimum Significant Difference= 0.1262

        Comparisons significant at the 0.05 level are indicated by '***'.

                         Simultaneous              Simultaneous
                            Lower     Difference      Upper
                 SLEEP    Confidence   Between     Confidence
                 Comparison  Limit      Means        Limit

             2    - 1       0.24375    0.37000      0.49625    ***
             1    - 2      -0.49625   -0.37000     -0.24375    ***
```

Review of Concepts

■ The SAS procedure for comparing more than two means is PROC ANOVA.

■ PROC ANOVA has two required statements: CLASS for identifying the categorical predictor variable, and MODEL for specifying the experimental design model you wish SAS to evaluate.

■ The output of PROC ANOVA is based upon regression analysis. The Summary Table is broken down into two parts, the first of which deals with the test of the overall model and the second of which deals with the test of the specific predictor variable(s) used. For simple one-way ANOVA such as we have been discussing, these two halves are equivalent.

■ The ANOVA printout also gives R^2 as a measure of association or size of effect.

■ PROC ANOVA can also take an optional MEANS statement, which can be used to request a printout of the Group means, or to request additional post-hoc analyses of comparisons between pairs of means.

■ Three procedures for conducting comparisons between pairs of means were discussed: Tukey's HSD (Tukey) and the Ryan Procedure (REGWQ) both of which rely upon the Studentized range statistic, and Fisher's LSD (LSD or T) which is simply a set of t-tests conducted under the "protection" of the significant ANOVA.

■ The statement Cldiff, changes the output from the Tukey and LSD procedures to report 95% confidence intervals on the differences between pairs of means rather than direct tests of the differences themselves.

■ PROC ANOVA can also perform factorial ANOVA tests. You simply list all the predictor variables in the CLASS statement, and then add terms for the main effects and interactions to the MODEL statement. Alternately you can just list the variable names separated by a vertical line in the model statement, if you want all main effects and interactions tested.

■ You can gain a better understanding of a significant interaction term by analyzing simple effects. You simply perform a one-way ANOVA using one of your predictor variables, BY the levels of the other predictor variable. Remember that you need to sort a data set by any variables used in subsequent BY statements.

Exercises

1. Use the data from **DS07_02.dat**. These data examining tranquilizer effects were used in the first worked example in Section 7.1 of the Text. Compare the means the three conditions using an one-way ANOVA. Further examine differences among the Groups using Tukey's HSD. Finally, have SPSS produce 95% confidence intervals for the Group differences, based upon the t statistic (Fisher's LSD).

2. Repeat the steps in Exercise 1 to the following data sets.

   a. **DS07_03.dat** (see Worked Example 2 in Section 7.1 of the text).

   b. **DS07_04.dat** (see Problem 1 in Section 7.1 of the text).

   c. **DS07_10.dat** (see Worked Example 2 in Section 7.3 of the text). This experiment requires a factorial ANOVA. Obtain a complete analysis of variance summary table and test the significance of main effects and the interaction.

5 MATCHED PAIRS AND WITHIN-SUBJECTS DESIGNS

Matched Pairs for Two Conditions

In Chapter 8 of the text a method was introduced for controlling extraneous sources of variance by matching subjects (or testing the same subjects) on the two conditions of an experiments. The net effect of this procedure was to reduce the variance attributable to residuals, when testing our model, by making the variance attributable to individual differences among subjects an explicit part of our model. To perform this operation for an experiment with two conditions, we first calculated difference scores for each subject, which were simply the subjects' score on the first condition minus the subjects' score on the second condition. We then tested the difference between the conditions by testing the null hypothesis that the mean of these difference scores was in fact 0.

To perform this procedure in SAS we will need to add a line to our data step to create the difference variable. Then we will have to find a way to test the mean of this newly created variable against the null-hypothesis that it is 0. Fortunately PROC UNIVARIATE, introduced in Section 2 of this workbook, can perform such a test for us. We will use the data presented in Section 8.2 of the text book, which presents a matched samples version of the experiment introduced in Chapter 6. The data for this experiment can be found in the file **DS08_01.dat**. A sample of the necessary program follows.

Example 1

```
OPTIONS ls=70 ps=54 nodate;

DATA Matched;

        INFILE "c:\datasets\DS08_01.dat";
        INPUT Pair Expmnt Control;

        Diff = Expmnt - Control;

RUN;

PROC UNIVARIATE;
        Var Diff;
RUN;
```

The line Diff = Expmnt - Control creates a new variable called Diff which contains the difference scores for the two conditions for each subject (or matched pair of subjects). We then call the UNIVARIATE procedure to produce its default list of descriptive statistics for this newly created variable. Note that we are not requesting the normal or plots options at this time, because we are primarily interested in one of the statistics that is produced by default. You should feel free, however, to embellish this procedure, using any of the techniques we discussed in Section 2, to more fully examine this variable. The results of running this program follow.

```
The SAS System

                         Univariate Procedure

Variable=DIFF


                              Moments

          N                 20    Sum Wgts              20
          Mean               4    Sum                   80
          Std Dev     7.426836    Variance        55.15789
          Skewness    0.287795    Kurtosis        -0.54934
          USS             1368    CSS                 1048
          CV          185.6709    Std Mean        1.660691
          T:Mean=0    2.408636    Pr>|T|            0.0263
          Num ^= 0          19    Num > 0             14
          M(Sign)          4.5    Pr>=|M|           0.0636
          Sgn Rank          50    Pr>=|S|           0.0431


                         Quantiles(Def=5)

          100%  Max          18        99%           18
           75%  Q3            7        95%           17
           50%  Med           3        90%           15
           25%  Q1           -1        10%           -6
            0%  Min          -8         5%           -7
                                       1%           -8
          Range             26
          Q3-Q1              8
          Mode               2


                             Extremes

          Lowest     Obs         Highest     Obs
              -8(       8)            8(       6)
              -6(      15)           14(       7)
              -6(       4)           14(      18)
              -4(      10)           16(       2)
              -2(       9)           18(      19)
```

This listing gives all the usual output we have come to expect from PROC UNIVARIATE. Check the calculated mean for the difference, for example, and compare it to that given in the text book. Of particular interest to us, however, is the line we have highlighted in the printout. The statistic T:Mean=0 2.408636 gives us the value of the paired samples *t*-test of the hypothesis that the mean of the difference scores is 0. The adjacent statistic, Pr>|T|0.0263, gives us the exact probability value associated with a *t* statistic of this magnitude, and these degrees of freedom. As can be seen, the value for the t score of approximately 2.41 agrees with the value calculated in the textbook, and the probability value of 0.0263 is certainly less than our usual alpha level of 0.05, thus we have reason to reject the null hypothesis. Remembering that the null hypothesis we are testing is that the mean of the differences between the conditions is 0, we can thus conclude that we have reason to believe there is a real difference between the conditions of our experiment. We will now examine the question of experiments with more than 2 conditions. As you might expect, we will have to shift our test-statistic of choice from t to F, as we more to more than 2 conditions.

Repeated Measures for More Than Two Conditions

The method for testing the null hypothesis that the means of more than two groups are the same, when we are sampling the same (or closely matched) subjects in each of our conditions may at first seem somewhat confusing. It is helpful to remember what we are attempting to do. In choosing to use a repeated measures methodology, we are attempting to reduce unaccounted for variance (the residual mean square) by making the variance due to individual differences between our subjects explicit in our model. Accordingly, even though we are only interested in one variable, our grouping variable, we will have to construct an ANOVA model as though we were interested in two, our grouping variable and the differences among the subjects. We accomplish this by using a variable containing an identifier of the subjects as part of our model, denoted in the text by π_j.

For our example we will be using the data from Section 8.3 of the text book. In this study, 17 subjects were tested in each of three conditions of testing time preceding and following the attribution manipulation. These data can be found in file: **DS08_07.dat**. The sample program for loading and testing this data set follows.

Example 2

```
OPTIONS ls=70 ps=54 nodate;

DATA Repeated;

        INFILE "c:\datasets\DS08_07.dat";
        INPUT Subject Attrib Score;

RUN;

PROC ANOVA;
        CLASS Attrib Subject;
        MODEL Score = Attrib Subject;
        MEANS Attrib / REGWQ;
RUN;
```

As you can see from the sample program, we are loading three variables into our data set. First we are loading a variable called subject which tells us which subject (or matched triplet of subjects) the data on that line is from. Next we have the variable Attrib which is our predictor variable of interest, and finally we have our response variable Score.

The ANOVA procedure should look fairly similar to examples from the last chapter, with two notable exceptions. As we mentioned, in order to eliminate the variability among subjects from our error term, we have to make it explicit in our model. Thus we list both Attrib and Subject in our Class statement, and in the right half of our model statement. Having produced similar factorial models in the last section, you might well ask why we have not included a third term in the model, representing the interaction of Attrib and Subject. The answer to this question is simple, with only 1 subject's score present in the cell at each combination of Attrib and Subject, there is no way to separate the variability due to the interaction from the error variability within cells. The two sources of variation are completely confounded. Our error term, *is* our interaction term. We conclude our ANOVA procedure with a means statement, using the Ryan procedure as a post-hoc test of the differences among our group means. The results from running this program are shown on page 204.

Note that the first output we are given is the listing of our two predictor variables, along with the levels of each. We should have three levels of Attrib and seventeen of Subject.

Next we have our ANOVA summary table. As we mentioned in the last section, SAS divides these tables into a top half dealing with the model as a whole, and a bottom half, dealing with each variable independently. From the top half of the table we can see that the error sum of squares for our model is 288.78 (note that this agrees with the results in Section 8.3 of the text book). Additionally we can see that the effect size of our overall model is $\eta^2 = R^2 = .627$. This tells us that our ANOVA model accounts for 62.7 percent of the total variability of the scores. Finally in the second half of the table we see the sums of squares, degrees of freedom and mean squares for both Attrib and Subjects. Note that these values also agree with those in Section 8.3. From the results we can conclude that there is reason to doubt the null hypothesis that the means of the three attribution conditions are equal. We usually don't care about the result of the test for subjects, though a significant result here, does indicate that we did a good job of partitioning off nuisance variability from our error term.

To examine the differences among our three means, we can now turn to the results of our post-hoc test(s). Note that although we have chosen to use the Ryan procedure again, you should feel free to use Tukey's HSD, or Fisher's LSD, or whichever test you prefer, simply refer to the previous section of the workbook for a reminder on how to invoke any of these other tests. The results for this test show that the means of Attrib condition 3 (Delayed Post test) and condition 2 (Immediate Post test) are both significantly greater than the mean of condition 1 (Pretest) but that they do not differ from each other. In other words the attribution manipulation worked, with evidence of learning being apparent immediately after the manipulation, and still apparent some time afterwards.

Review of Concepts

■ Using matched samples, or the same subjects are methods of reducing the variance in our models that we have to attribute to error, by making the individual differences between subjects explicit.

■ When you have an experiment with two conditions the test of choice is a *t*-test performed on difference scores calculated between our conditions.

■ To perform such a test in SAS you first have to create the new response measure during the DATA step. You then use the UNIVARIATE procedure to furnish descriptive statistics for this difference variable. One of the default descriptives produced is a *t*-test of the null hypothesis that the mean of a variable is zero. We can use this as our test for differences between our group means.

```
Analysis of Variance Procedure
                   Class Level Information

      Class   Levels   Values

      ATTRIB     3     1 2 3

      SUBJECT   17     1 2 3 4 5 6 7 8 9 10 11 12 13 14 15 16 17

              Number of observations in data set = 51

                      The SAS System

                 Analysis of Variance Procedure

Dependent Variable: SCORE

Source                  DF    Sum of Squares    F Value      Pr > F

Model                   18     485.96078431       2.99       0.0033

Error                   32     288.78431373

Corrected Total         50     774.74509804

              R-Square              C.V.            SCORE Mean

              0.627252           17.17581          17.4901961


Source                  DF       Anova SS        F Value      Pr > F

ATTRIB                   2     127.21568627        7.05       0.0029
SUBJECT                 16     358.74509804        2.48       0.0139


  Ryan-Einot-Gabriel-Welsch Multiple Range Test for variable: SCORE

     NOTE: This test controls the type I experimentwise error rate.

              Alpha= 0.05  df= 32  MSE= 9.02451

         Number of Means          2          3
         Critical Range   2.098852  2.5320553

  Means with the same letter are not significantly different.

          REGWQ Grouping          Mean     N  ATTRIB

                       A         18.941    17   3
                       A
                       A         18.235    17   2

                       B         15.294    17   1
```

■ In the event that we have more than two conditions in an experiment we will rely on a repeated measures ANOVA as our test.

■ Performing such a test in SAS is virtually identical to the one way ANOVA we discussed in the last section. You just need a variable coding the identity of subjects in you data set and then you add this term for subjects to both you class and model statements. You then examine the results of your response variable exactly as you would have in Section 4.

Exercises

From Chapter 8 of the textbook, try to use SAS to answer the questions in Exercise 5 and Problems 1 and 2. In addition to the questions asked, perform a post-hoc test (of your choice) on the differences between the group means in each problem.

6 REGRESSION AND CORRELATION

Linear Regression

In Chapter 9 of the textbook, you were introduced to a statistical technique called linear regression. Unlike the techniques described in the previous chapters, regression has the advantage of allowing you to develop a model describing a functional relationship between two (or more) variables. In the textbook this equation is written as $\hat{Y} = \alpha + \beta X$, where α is the intercept and β is the slope. The methodology you learned in Chapter 9 obtains estimates (α and β) of the parameters a and b. On the basis of these values you construct an equation from which you can predict the values of Y by multiplying the estimated slope by the values of X and then adding estimated value of the intercept. We can test how well our line fits the data, first by calculating sums of squares for our regression line, and then for the residuals (differences between our predicted values of Y and our obtained values of Y). The result is an ANOVA table testing the null hypothesis that $\beta = 0$.

We can further examine our residuals by plotting them against values of our response measure and against our predicted values to look for any systematic deviations from the straight line our model predicts. As an example of this methodology we will be using the first data set found in Section 9.1 of the textbook. These data can be found in file **DS09_01.dat**. Our response variable is the number of errors made by subjects under varying conditions of sleep deprivation, while our predictor variable is the number of hours of sleep deprivation experienced by that subject. In other words our Y variable is Errors, while our X variable is Hours. We will now construct a model specifying a linear relationship between these variables and then test that model using an ANOVA, and using graphical analyses of the residuals from our model. The program to perform this analysis follows.

```
OPTIONS ps=54 ls=70 nodate;

DATA Sleep;
        INFILE 'c:\datasets\DS09_01.dat';
        INPUT Hours Errors;
RUN;

PROC REG;
        MODEL Errors = Hours;
        Plot Predicted.*Errors
            Student.*Predicted. / Vplots = 2;
RUN;
```

The DATA step should be fairly old hat for you by now. The SAS procedure for performing a regression analysis is, quite naturally, PROC REG. This is one of the largest, most elaborated procedures in SAS. We will endeavour to keep our example simple.

The one required statement in PROC REG is the MODEL statement. This operates much the way that the MODEL statement in PROC ANOVA did. This should come as no surprise, given that analysis of variance is really just a special case of regression analysis. Indeed the SAS procedures for regression and ANOVA were both based on the original procedure for performing both operations called PROC GLM (for general linear model). After specifying our model in the form response = predictor (or dependent = independent), I then added a statement to produce two plots which will help us examine the residuals of our model. The optional Plot statement in PROC REG takes the form: Y axis * X axis / options. You can request multiple plots with the same command simply by leaving a space between them (remember SAS thinks its reading the same line until it reaches a ;). You then follow with the slash and any options you wish. Here we have invoked the Vplots option, this tells SAS how many plots it should fit vertically on a page. (Obviously there is an analogous Hplots option). For the variable names in the Plot statement you can use any variable in your data set, or some that SAS will create in the course of running the regression. We have used two of the latter: Predicted and Student. These are SAS command words for predicted values and Studentized (t-transformed) residuals, respectively.

```
The SAS System

Model: MODEL1
Dependent Variable: ERRORS

                    Analysis of Variance

                        Sum of         Mean
Source          DF     Squares       Square     F Value     Prob>F

Model            1   2883.69000   2883.69000     70.269     0.0001
Error           48   1969.83000     41.03813
C Total         49   4853.52000

     Root MSE        6.40610     R-square       0.5941
     Dep Mean       29.64000     Adj R-sq       0.5857
     C.V.           21.61302

                    Parameter Estimates

                 Parameter     Standard    T for H0:
Variable   DF     Estimate       Error    Parameter=0    Prob > |T|

INTERCEP    1    13.530000    2.12466321      6.368        0.0001
HOURS       1     0.671250    0.08007626      8.383        0.0001
```

The output from PROC REG can be roughly divided into two halves, the first dealing with the model as a whole. It is here that we get our value of R-square (R^2) for our model, along with a standard error on this estimated value, enabling us to place confidence intervals around the estimate. R-square is our measure of association (or alternately of effect size). Our value of 0.5941 means that we can account for approximately 60% of the variance in subjects error score based upon our knowledge of the number of hours they have been deprived of sleep. It thus presents an index of the predictive power of our model. Note that there is an additional value called Adjusted R-Square.

In general as you use more predictor variables in your model, the value of R-square will become positively biased, purely as an artifact of the way we calculate our model sums of squares. Accordingly it becomes increasingly desirable to have an estimate of degree of association that is unbiased as we add more predictors. Note that with only the one predictor variable there is virtually no difference between the two values.

Next we have the ANOVA summary table for our model. Note that the sums of squares, degrees of freedom, etc. correspond to those in the text book. Our regression has an F-value > 70, which in turn we are told is significant at a p-value <0.001. We can thus safely discredit the null hypothesis that our coefficients = 0 and thus have one more reason to believe that our model fits the data well.

The second half of the output is the table listing our model coefficients. According to this table our best model for predicting errors is: errors = 0.67125 (hours) + 13.53. In addition to the point estimates of our slope and intercept we are given a standard error on each estimate. This allows us to compute a 95% confidence interval on each coefficient. Additionally there are the two t-tests, each testing the null hypothesis that that particular coefficient is 0. As can be seen we have reason to discredit both of these lesser null hypotheses as well as the overall null hypothesis tested by the model ANOVA.

Finally, we can examine our two plots for any problems with our model. You will recall that the first chart we requested was a plot of our predicted values against our response measure. Second we requested a plot of our Studentized residuals against our predicted values. The charts we selected can be seen below. In the first chart we are looking for any deviation from a linear relationship. If there were any curves apparent in this chart it would be evidence that the best model for our data was probably not a linear model. At that point we would have several options, the most commonly exercised being a return to the data set, transforming either or both variables in an attempt to linearize their relationship and then a new regression analysis on the transformed data. The second chart serves a somewhat similar function. It is a test of the adequacy of our model across its range. The residuals should be dispersed fairly evenly to either side of the zero line, along the length of the x axis. If there were any patterns evident in their dispersal, we would again have evidence of problems with our model and would probably have to adopt a strategy similar to that for problems with the first chart (see page 209).

SAS uses numbers as its plotting symbols. This enables you to see overlapping data points as progressively higher numbers. (In other words if there were three data points at one location on the graph SAS would plot it as a 3). If you find this confusing, you can instruct SAS to use another plotting symbol when you request the plots. You simply add an equal sign to the request, such as Plot Predicted.*Student.='*', which would use * as the plotting symbol on the charts, with the loss of information about number of data points at a given location. As can be seen, there are no serious problems with either chart, and thus we can probably safely consider our model a good one. We now have three pieces of evidence, the value of R-square, the significant ANOVA and the residual plots indicating a good model has been selected.

Pearson's Correlation Coefficient

Occasionally we are not interested in predicting one variable on the basis of our knowledge of another variable. We merely wish some index of how closely the two variables are related to each other. The statistic of choice in such cases is Pearson's product-moment correlation coefficient, r. As stated in the text, this statistic is very popular, though it should probably always be kept in mind that the underlying model of r is the same as that of regression (which is why the symbol is an r to begin with), namely that there is a linear, functional relationship between your two variables. As a regression analysis is a more fully elaborated technique, allowing for many nuances of application,

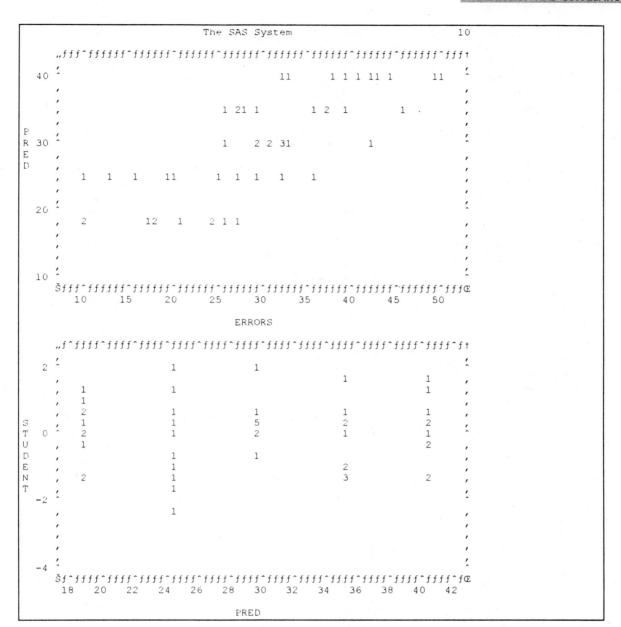

and many diagnostic techniques, it is often to be preferred over *r*. However, occasionally you will just want a correlation between two variables without thought of direction, and thus we will show you how to calculate one. The example we will use is the third data set from Section 9.1 of the textbook, the relationship between IQ scores and GPA scores for 50 subjects. These data can be found in file **DS09_03.dat** and we have labeled the two variables IQ and GPA respectively. To request a correlation from SAS the program at the top of page 210 will be effective.

PROC CORR is the procedure for calculating correlation coefficients. By default it will calculate a Pearson's *r* for every pair of variables in your data set. You can request different correlation coefficients (such as Spearman's rho or Kendall's tau) through the use of options on the procedure line (Spearman, Kendall, Pearson, etc.). If you specify one of these other correlations, you will not be given the default Pearson's *r* unless you request it specifically. Finally, you can limit the variables you want correlated with one another using an optional Var statement. The results of this program are shown in the middle of page 210.

```
OPTIONS ps=54 ls=70 nodate;

DATA Iqgpa;

        INFILE 'c:\datasets\DS09_03.dat';
        INPUT IQ GPA;

RUN;

PROC CORR;
        Var IQ GPA;
RUN;
```

```
The SAS System

                        Correlation Analysis

            2 'VAR' Variables:   IQ          GPA


                        Simple Statistics

Variable        N       Mean    Std Dev      Sum   Minimum   Maximum

IQ             50      110.2    11.7902   5509.0   83.0000    136.0
GPA            50     2.2640    1.0781    113.2    0.1000    4.0000


Pearson Correlation Coefficients / Prob > |R| under Ho: Rho=0 / N = 50


                                IQ              GPA

            IQ              1.00000          0.53436
                              0.0              0.0001

            GPA             0.53436          1.00000
                              0.0001           0.0
```

As can be seen, the value of our correlation coefficient is 0.534. You can square this result to get a measure of effect size. $(0.534)^2 = 0.285$, so we can account for 28.5% of the variation in a persons GPA score by knowing their IQ (and vice versa). The probability values associated with the *t*-test are less than 0.01 according to our output, and thus we have reason to discredit the null hypothesis that the actual relationship between these scores is 0. We thus have reason to believe that there is a moderate relationship between IQ and GPA.

Review of Concepts

■ Regression analysis is a technique which allows us to generate models of a functional, linear relationship between two (or more) variables. The form of the model is given by the slope and intercept of the straight line that best describes this relationship.

■ One can test the fit of a regression model in (at least) three ways. One can examine the size of effect of the model, given by the model's R-square. One can examine an ANOVA which tests the null hypothesis that our coefficients are all zero. Finally, one can examine the residuals of our model (the predicted values minus our original response values) graphically.

■ To perform a regression analysis in SAS one uses PROC REG. It has a mandatory MODEL statement which works like the MODEL statement in PROC ANOVA. You can request various plots using the optional Plots statement in PROC REG.

■ Occasionally one is interested in a nondirectional (and thus nonpredictive) measure of association between two variables. The statistic of choice in this situation is Pearson's correlation coefficient, r. It can be evaluated by squaring it to give a magnitude of effect measure. SAS also converts it to a t statistic, testing the null hypothesis that the actual relationship between the two variables is zero.

■ To request a correlation between two variables you use PROC CORR. It can perform other correlations than Pearson's by request of an option on the procedure line. The Var statement limits this procedure as many others to a subset of our available variables.

Exercises

1. Example 2 in Section 9.1 examined test scores as a function of age. Using the file **DS09_02.dat**, obtain a linear regression summary table, estimates of α and β, and a 95% confidence interval for β.

2. Problem 4 of Section 9.1 investigates the relation between maternal attention and infant attractiveness. Using the file **DS09_09.dat**, obtain the Pearson product-moment correlation coefficient between these two variables.

3. Problem 1 of Section 9.2 investigates the relation between memory and marijuana. Using the file **DS09_06.dat**, obtain a linear regression summary table, estimates of α and β, and a 95% confidence interval for β.

   Using files **DS09_13.dat**, **DS09_14.dat**, and **DS09_15.dat**, obtain Pearson product-moment correlation coefficients for the IQs of fraternal twins reared together, identical twins reared apart, and identical twins reared together (Chapter 9, Exercise 2).

7 ANALYSIS OF CONTINGENCY TABLES

Goodness of Fit

In chapter 10 of the textbook you were introduced to the analysis of categorical data. While we have already used one categorical variable in both independent samples *t*-tests and analysis of variance, both of those procedures had response variables that were continuous. When we have two (or more) variables that are categorical and we wish to test the frequency of subjects in each of those categories, we must rely on other procedures. The simplest of these is the goodness of fit test, which has only one variable, and the frequencies of subjects in the different levels of that variable.

In SAS when we want to look at tables of frequencies we use the command PROC FREQ, with a TABLES statement, as introduced in Section 2 of this workbook. For our example of a one-way table suitable for a goodness-of-fit test, we use one of the examples from Chapter 10 of the text. This is the experiment in which the researchers ask each student in a class to think of a single digit ranging from 0 to 9. The data for this experiment is not on the disk, but as you will see, it is easy enough to enter directly. The program follows.

```
OPTIONS ls=70 ps=54 nodate;

DATA Goodfit;

        INPUT Digit Count;

LINES;
0       13
1       16
2       19
3       51
4       26
5       28
6       37
7       43
8       22
9       15
RUN;

PROC FREQ;
        TABLES Digit;
        WEIGHT Count;
RUN;
```

As you can see we did not try to recreate the raw data in the form of one line per subject, indicating which digit they had thought of. Instead we included a second variable called count, which stored the frequency counts of subjects imagining each digit. To tell SAS to use this variable we have to add an optional WEIGHT statement to PROC FREQ. This statement tells SAS not to try and count up the frequencies of Digit itself, but instead to look for those frequencies in our variable Count. This notation can save a lot of space in data entry. The results for running this program follow.

```
The SAS System

                                      Cumulative   Cumulative
        DIGIT   Frequency    Percent   Frequency    Percent
ffffffffffffffffffffffffffffffffffffffffffffffffffffffff
          0          13        4.8          13         4.8
          1          16        5.9          29        10.7
          2          19        7.0          48        17.8
          3          51       18.9          99        36.7
          4          26        9.6         125        46.3
          5          28       10.4         153        56.7
          6          37       13.7         190        70.4
          7          43       15.9         233        86.3
          8          22        8.1         255        94.4
          9          15        5.6         270       100.0
```

As you can see, SAS has correctly tabulated our frequencies, and then calculated cumulative frequencies and percentages for each digit imagined. Unfortunately, SAS can do no more with a one way table of frequencies. Strange as it may seem there is simply no easy way to get SAS to compute a χ^2 test of goodness of fit, with only one variable in PROC FREQ. Hopefully tabulating the results will at least facilitate manual calculation of the test. Feel free to write to the SAS Institute to inquire about this puzzling oversight. The situation improves greatly once we move on to contingency tables (tables with more than one variable) such as the classic 2×2 contingency table for which χ^2 is the test statistic of choice.

2×2 Contingency Tables

For our example of a 2×2 contingency table, we will again lift an example directly from the pages of the textbook. In this experiment subjects were asked to state a preference for photos, when the available choices were either a photo of themselves, or a photo of a good friend, when either photo could be displayed in either normal or mirror image orientation. It was predicted that subjects should prefer the normal orientation of their friend's face but the mirror image orientation of their own face, both corresponding to their usual experience. The program to read this data set follows.

```
OPTIONS ps=54 ls=70 nodate;

DATA Twobytwo;

        INPUT Photo $ Orient $ Count;

Lines;
Friend  Original        21
Friend  Mirror           4
Self    Original         9
Self    Mirror          16
RUN;

PROC FREQ;
        TABLES Photo * Orient / Expected Chisq;
        WEIGHT Count;
RUN;
```

We specify the 2 × 2 table by listing the two variables joined by an asterisk. As you can see, now that we have a table of greater than one dimension we can add optional statements to our TABLES statement. The Expected statement requests that expected values for each cell be printed in the output, while the Chisq statement requests a direct test of the null hypothesis of independence using the χ^2 statistic, as well as producing several other measures of association. (Note, you *can* add these statements in the goodness-of-fit case above, they just won't do anything.) The results of this program are as follows.

```
The SAS System
                    TABLE OF PHOTO BY ORIENT

        PHOTO          ORIENT

        Frequency,
        Expected ,
        Percent  ,
        Row Pct  ,
        Col Pct  ,Mirror  ,Original, Total
        ffffffffff^ffffffff^ffffffff^
        Friend   ,      4 ,     21 ,    25
                 ,     10 ,     15 ,
                 ,   8.00 ,  42.00 , 50.00
                 ,  16.00 ,  84.00 ,
                 ,  20.00 ,  70.00 ,
        ffffffffff^ffffffff^ffffffff^
        Self     ,     16 ,      9 ,    25
                 ,     10 ,     15 ,
                 ,  32.00 ,  18.00 , 50.00
                 ,  64.00 ,  36.00 ,
                 ,  80.00 ,  30.00 ,
        ffffffffff^ffffffff^ffffffff^
        Total           20       30       50
                     40.00    60.00   100.00

        STATISTICS FOR TABLE OF PHOTO BY ORIENT

Statistic                      DF    Value      Prob
ffffffffffffffffffffffffffffffffffffffffffffffffffff
Chi-Square                      1    12.000     0.001
Likelihood Ratio Chi-Square     1    12.647     0.001
Continuity Adj. Chi-Square      1    10.083     0.001
Mantel-Haenszel Chi-Square      1    11.760     0.001
Fisher's Exact Test (Left)                    6.04E-04
                    (Right)                    1.000
                    (2-Tail)                  1.21E-03
Phi Coefficient                     -0.490
Contingency Coefficient              0.440
Cramer's V                          -0.490
        Sample Size = 50
```

As you can see, SAS will produce a complete contingency table for us, with row and column totals as well as expected frequencies in each cell. The second half of the output is a table of statistical tests. Of primary interest to us here are the value labeled chi-square which for these data is $\chi^2(1)=12.00$, which we are informed is significant at below 0.001. As this result is far below our normal alpha level of 0.05 we have statistical justification for discrediting the null hypothesis that the cells are independent. In other words the cells are sufficiently unequal in frequency for us to feel justified in claiming that there is a real phenomenon (in fact the one we predicted) present. Also of interest is the phi coefficient. As stated in the textbook, this is a measure of effect size, akin to Pearson's correlation coefficient. Its absolute value is 0.49, which is quite respectable, particularly in the psychology literature.

Larger Contingency Tables

The methodology employed above extends directly to cases with more than 2 rows and/or columns. For an example of a larger contingency table we use a data file corresponding to Exercise 2 in Chapter 10, file **DS10_01.dat**. The data are in raw form. In other words, each line of data corresponds to one subject, with the two variable indicating which of the three groups they are members of and which of the two responses they made, respectively. We can thus let SAS count up our frequencies for us and will not require a WEIGHT statement in our FREQ procedure. The program to load and analyze these data follows.

```
OPTIONS ps=54 ls=70 nodate;

DATA RbyC;

        INFILE 'c:\datasets\DS10_01.dat';
        INPUT Group Resp $;
RUN;

PROC FREQ;
        TABLES Group * Resp / Expected Chisq;
RUN;
```

The syntax is identical to the 2×2 case. The results appear on page 216.

Once again the output is divided into two halves. The contingency table is presented first, followed by the table of statistics. Our χ^2 is once again significant with a value of 10.069 at 2 degrees of freedom. It should be noted that when we are looking at tables with more than 2 rows and columns, however, that we have to make an adjustment to our calculation of phi. As stated in the text book, this adjustment was proposed by Cramer, and the statistic is thus commonly known as Cramer's phi, which is somewhat inexplicably rendered as Cramer's V in the SAS output. Presumably it was felt that a V was the closest ASCII character available to a phi, but that still doesn't explain why the statistic wasn't simply labeled Cramer's phi. Whatever the explanation, we have a value of 0.305 for this statistic which is still of moderate size, though less than that of our previous example.

Review of Concepts

■ When we wish to analyze data in the form of frequency counts, the methodology involves the construction of contingency tables and the subsequent testing of observed frequencies against the expected values of frequencies under the null hypothesis of independence.

■ In SAS these tests are conducted using PROC FREQ with a TABLES statement.

■ In the case of one-way contingency tables, or goodness of fit tests, SAS cannot calculate the relevant test statistic, but the syntax for creating the contingency table remains the same.

```
TABLE OF GROUP BY RESP

            GROUP       RESP

            Frequency,
            Expected ,
            Percent  ,
            Row Pct  ,
            Col Pct  ,A        ,R        ,  Total
            ƒƒƒƒƒƒƒƒƒ^ƒƒƒƒƒƒƒƒ^ƒƒƒƒƒƒƒƒ^
               1 ,       8 ,      28 ,        36
                 , 12.667 , 23.333 ,
                 ,   7.41 ,  25.93 ,    33.33
                 ,  22.22 ,  77.78 ,
                 ,  21.05 ,  40.00 ,
            ƒƒƒƒƒƒƒƒƒ^ƒƒƒƒƒƒƒƒ^ƒƒƒƒƒƒƒƒ^
               2 ,      20 ,      16 ,        36
                 , 12.667 , 23.333 ,
                 ,  18.52 ,  14.81 ,    33.33
                 ,  55.56 ,  44.44 ,
                 ,  52.63 ,  22.86 ,
            ƒƒƒƒƒƒƒƒƒ^ƒƒƒƒƒƒƒƒ^ƒƒƒƒƒƒƒƒ^
               3 ,      10 ,      26 ,        36
                 , 12.667 , 23.333 ,
                 ,   9.26 ,  24.07 ,    33.33
                 ,  27.78 ,  72.22 ,
                 ,  26.32 ,  37.14 ,
            ƒƒƒƒƒƒƒƒƒ^ƒƒƒƒƒƒƒƒ^ƒƒƒƒƒƒƒƒ^
            Total        38       70       108
                      35.19    64.81    100.00


                    The SAS System

        STATISTICS FOR TABLE OF GROUP BY RESP

    Statistic                   DF    Value       Prob
    ƒƒƒƒƒƒƒƒƒƒƒƒƒƒƒƒƒƒƒƒƒƒƒƒƒƒƒƒƒƒƒƒƒƒƒƒƒƒƒƒƒƒƒƒƒƒƒƒƒƒ
    Chi-Square                   2   10.069      0.007
    Likelihood Ratio Chi-Square  2    9.954      0.007
    Mantel-Haenszel Chi-Square   1    0.241      0.623
    Phi Coefficient                   0.305
    Contingency Coefficient           0.292
    Cramer's V                        0.305

    Sample Size = 108
```

■ For all contingency tables with more than one variable, you can add the options Expected and Chisq to the Tables statement to calculate expected frequencies and test the null hypothesis of independence.

■ In addition, SAS prints measures of association including the phi coefficient, and Cramer's phi (labeled V). These can be interpreted as measures of effect size analogous to Pearson's r.

Exercises

Answer the questions in Exercises 3 and 4 of Chapter 10 using SAS.

Appendix A Answers to Problems and Exercises
CHAPTER 1 Purpose of Statistical Data Analysis
Short-Answer Questions

1. See Section 1.1 of textbook.

2. John's height. Height is less variable than weight.

3. b, a, c. Proposal b invokes two causes, a invokes one, and c none.

4. Possible examples are (a) categorical: employment status, job classification; (b) quantitative: age, IQ.

5. The child chosen randomly from the Grade 2 class. The variability would be less than in the more general group.

6. A speedometer that was biased might show a speed that was always 5 mph slower than the actual speed; one that lacked precision would give values that fluctuate even when the car was travelling at a constant speed.

Short Problems

Problem 1. The predictor is annual income (quantitative, natural), the response variable is stated voting intention (categorical).

Problem 2. The predictor is GPA (quantitative, natural), the response variable is stated painting preference (categorical).

Problem 3. (i) c, f ; (ii) b, e, h; (iii) a, g; (iv) none; (v) g; (vi) a, d

Problem 4. (i) c, f; (ii) b, e, h; (iii) a, g; (iv) none; (v) g; (vi) a, d

Exercises in the Analysis of Realistic Data

Exercise 1. a. The predictor variable is study condition (read versus generate) which is categorical, manipulated. The response variable is recall score which is quantitative b. Yes, bias. c. Within each condition they are variable. The most likely sources of noise for such data is individual differences in memory ability d. less.

Exercise 2. a. The predictor variable is Group (expert versus novice) which is categorical, natural. The response variable is recall score which is quantitative b. Perhaps chess experts had better memories to start with. Natural predictor variable.

Exercise 3. a. The predictor variable is person making request (parent versus stranger) which is categorical, manipulated. The response variable is time to comply which is quantitative b. (ii) is more plausible. The random assignment makes (i) an unlikely explanation.

Exercise 4. a. The predictor variable is mother's authoritarianism score which is quantitative, natural. The response variable is average course grade which is quantitative b. The predictor variable being natural makes this causal conclusion questionable.

Exercise 5. a. The predictor variable in mood (good versus neutral) which is categorical, manipulated. The response variable is time to comply which is quantitative b. Random assignment of children to conditions.

CHAPTER 2 Graphical and Numerical Descriptions of Data

Short-Answer Questions

1. a. In a histogram there is no space between the bars b. A bar chart is used for categorical variables; the histogram for quantitative variables. c. In a bar chart the ordering is arbitrary.

2. See Section 2.1.3 of the textbook.

3. See Section 2.1.4 of the textbook.

4. In a bar chart the ordering of the values on the x-axis is arbitrary.

5. See Section 2.2.1 of the textbook. A quintile would divide the distribution into five equal areas.

6. See Section 2.2.3 of the textbook. Yes, the IQR is based on quartiles which, like the median, are resistant statistics.

7. See Section 2.2.1 of the textbook.

8. The variance is the average of the sum of squared deviations of the scores about their mean.

9. See Section 2.2.4 of the textbook.

10. The mean is 0 and the variance is 1.

11. Yes for both.

12. Rescaling scores cannot change their rank order.

Short Problems

Problem 1. Mean = 170 cm, the median is 167.5 cm and the standard deviation = 8.75 cm.

Problem 2. The mean. The high-end outliers will make the mean greater than the median.

Problem 3. Her z-score is 0.563; the rescaled score = .563x10 + 70 = 76.

Problem 4. Positive skew or high-end outliers. Bimodality.

Problem 5. a. F; b. F; c. T; d. T; e. F; f. T; g. F; h. F; i. T.

Exercises in the Analysis of Realistic Data

Exercise 1. 3. The stemplot for the 3-second and 6-second retention interval indicates negative skew whereas the 18-sec retention interval indicates positive skew. The negative skew reflects a ceiling effect (very easy condition); the positive skew a floor effect (very difficult condition).

Exercise 2

1.

Read		Generate	
5	0	0	99
6	0	1	00
7	0000	1	233
8	000	1	4444445
9	000	1	6666677
10	00000	1	8899
11	00		
12	000		
13	00		
14	0		

3.

	n	Var.	SD	*IQR*
Read	25	5.59	2.36	4.0
Generate	25	8.43	2.90	3.5

4. No

Exercise 3

1.

Experts		Novices	
0	4	0	344
0	667899	0	5555667778999
1	012222334	1	001133
1	56678		
2	1		

2.

	Var.	SD
Experts	18.50	4.30
Novices	8.54	2.92

3. No

5. The predictor variable in this experiment is natural. It makes a causal interpretation difficult. Refer to Exercise 2 in Chapter 1.

Exercise 4:

1.

Expert		Novice	
3	0	2	0
4	00	3	000
5	000	4	0
6	00	5	00
7	00	6	0000
8	0000	7	000
9	000000000	8	00000000
10	000	9	00
11	00	10	00
		11	00

	SS	SD	IQR
Experts	131.22	2.20	3.0
Novices	159.84	2.43	2.75

3. No

Exercise 5

1. Construct and examine a stemplot, calculate the *IQR* and check for outliers.

3. The mean of 19.9 is for the stranger condition, the mean of 30.0 is for the parent condition

4. For the parent group the standard deviation is 9.78 and for the stranger-request group it is 8.00.

Exercise 6

1. The means are 2.6, 3.9, 4.0 for conditions 1, 2, and 3 respectively.

2. the standard deviations are 1.35, 1.16, 1.25 for conditions 1, 2, and 3 respectively.

Exercise 7

2. Mother's authoritarianism, *IQR* = 12.5; Child's school grade, *IQR* = 11.5.

3. No

Exercise 8

3.

	SD	IQR
Family routine chores	8.85	10.5
Concern-for-others	7.85	10.5

4. Calculation of *z*-scores would show that both scores are above their respective means. However, this is an observation for just a single child and provides no evidence that this positive relationship holds more generally.

Exercise 9

2.
Var.	SD	IQR
808.35	28.43	39

3. No

5. The tabled entries be for original scores are: 70: 44; 80: 48; 90: 51.

Exercise 10

1.
```
2 68
3 024
3 67779
4 001333334444
4 555677999
5 0011223344444
5 555566667778888
6 00000112222223333444
6 5555555666778999
7 11233344
7 555679
8 123344
8 6779
9 3
```

2. $Q_1 = 49$ $Q_2 = 67$

3. $IQR = 18.00$

CHAPTER 3
Modeling Data and the Estimation of Parameters
Short-Answer Questions

1. $\hat{Y}$ is the predicted value whereas Y is the observed value of the response variable.

2. See Section 3.1.2 of textbook.

3. μ can be thought of as analogous to a signal and e can be thought of as analogous to noise.

4. Because their values are unknown.

5. The claim has a mathematical proof.

6. The value of SS_e must always be positive because it is a sum of squared values.

7. Because it is the ratio of sums of squares which must be positive.

8. See Section 3.3.1 of textbook.

9. $SS_{total} = SS_{model} + SS_e$.

10. Both are expressed in units of the standard deviation.

Short Problems

Problem 1. $SS_e = 20.0$; $MS_e = 5.0$

Problem 2. $SS_e = 18.0$

Problem 3. $SS_e = 38$ and $MS_e = 4.75$

Problem 4. $SS_{total} = 48$

Problem 5. $R^2 = .208$, $d = .918$.

Problem 6. The value of SS_e is unchanged because adding the constant does not change the variability within a condition. The new values are $SS_{total} = 78.0$, $d = 1.835$, and $R^2 = .513$

Exercises in the Analysis of Realistic Data

Exercise 1
1. The estimate of μ_1 is 9.5 and of μ_2 is 14.5.

2. $SS_e = 336.48$; $MS_e = 7.01$

3. $\hat{d} = 1.89$.

4. $Y = \mu + e$. $SS_{model} = 648.98 - 336.48 = 312.50$. $R^2 = 312.50/648.98 = .48$.

Exercise 2
1. The full model is $Y_1 = \mu_1 + e$; $Y_2 = \mu_2 + e$. The null model $Y = \mu + e$. SS_e 567.91, $MS_e = 13.52$.

2. $SS_{model} = 200.82$ and $R^2 = .261$.

Exercise 3

$\hat{d} = 10.1/8.93 = 1.13$; $R^2 = .25$

Exercise 4

1. The full model is $Y_1 = \mu_1 + e$; $Y_2 = \mu_2 + e$; $Y_3 = \mu_3 + e$. The null model $Y_1 = Y_2 = Y_3 = \mu + e$.

2. For the full model, the estimate of μ_1 is 2.6, of μ_2 is 3.9, and of μ_3 is 4.0. For the null model, the estimate of μ is 3.5.

3. $SS_e = 66.53$; $MS_e = 1.58$

4. $SS_{model} = 18.71$ and $R^2 = .22$.

CHAPTER 4 Probability Distributions
Short-Answer Questions

1. The smaller sample is more likely than the larger sample to have a majority of women because in this population women are a minority. According to the law of large numbers, a departure from this population property is more likely with a small sample.

2. See Section 4.1.4 in textbook.

3. A binomial distribution with $n = 9$ and $p = 1/3$. Note that the numbering of the bins corresponds to the number of right deflections, thus the probability is $p = 1/3$, not $2/3$.

4. The probabilities remain 1/3 right, 2/3 left for all falls, regardless of previous outcomes. Given that each player has a certain overall probability of success, this probability is not influenced by the outcome of the previous toss.

5. This measurement is a random variable in the sense that probability statements could be made about its values. For example it might be stated that the probability that this person's reaction time exceeds 300 milliseconds is .37. IQ could be considered a random variable if values are considered across individuals so that claims could be made such as the probability that a randomly selected person's IQ exceeds 110 is .25.

6. See Section 4.3.3 in textbook.

7. Physical features such as chest circumference are assumed to be the result of a large number of additive independent influences.

8. Symmetry makes it possible to read off values for negative scores from the values for positive scores.

Short Problems

Problem 1

0	1	2	3	4	5	6	7	8	9
.026	.117	.234	.273	.205	.102	.034	.007	.001	.000

The probability that the next ball will also land in bin 4 is .205.

Problem 2. The probability is .114

Problem 3. Top 5%: $z = 60 + 10 \times 1.64 = 76.4$; bottom 10%: $z = 60 - 10 \times 1.28 = 47.2$

Problem 4. 81 and 119.

Problem 5. .363 .144

Problem 6. a. $Q_1 = 90.0$, $Q_2 = 110.0$. b. IQ = 119.0. c. .9082. d. .4082. e. 9.5%

Problem 7. $z = \pm 1.64$. $50 \pm 1.64 \times 13$. That is, 28.7 and 71.3.

Problem 8. $70 \pm .67 \times 14$. That is, 60.6 and 79.4.

Exercises in the Analysis of Realistic Data

Exercise 1. The z-scores for quintiles in a normal distribution are –0.84, –0.25, +0.25, and +0.84. In a distribution with a mean of 86.35 and a standard deviation of 28.43, the quintiles (Quin) are $Quin_1$ = 86.35 – .84 × 28.43 = 62.5; $Quin_2$ = 86.35 – .25 × 28.43 = 79.2; $Quin_3$ = 86.35 + .25 × 28.43 = 93.5 Quin4 = 86.35 + .84 × 28.43 = 110.2. Thus (a) 60 is E, (b) 85 is C, (c) 100 is B, (d) 120 is A..

Exercise 2. The z-scores corresponding to the two boundaries are –0.43 and +0.43. The two boundaries are 52.9, and 65.1. (a) 50 is Introvert; (b) 60 is Intermediate, and (c) 70 is Extrovert.

CHAPTER 5 Sampling Distributions and Interval Estimation

Short-Answer Questions

1. See Section 5.1.1 in textbook. (a) The population from which this "sample of size one" has been drawn is the set of all sample means from all samples of size 10. (b) The form of the probability distribution from which it was drawn is the normal distribution.
2. You should agree. The standard error is smaller for the larger sample.
3. The sampling distribution of the mean will be normal.
4. A point estimate consists of a single value, an interval estimate consists of a range of values.
5. The interval for the smaller sample would be wider. The standard error for the smaller sample would be larger. With a smaller sample, a wider interval is needed to achieve the same level of confidence.
6. A 99% confidence interval.
7. Use the t-distribution when the population standard deviation is unknown and must be estimated. Use the normal distribution when it is known.
8. Narrower. z from the normal distribution is always less than the corresponding value of t.
9. The second. The larger variance would result in a larger standard error and thus a wider interval.
10. No, the predicted value of 5.0 is not a plausible one. It lies outside the confidence interval.

Short Problems

Problem 1. $\sigma_{\bar{Y}} = 4.69$; $z = (48 - 50)/4.69 = -0.43$. Prob. $= .6664$.

Problem 2. $\sigma_{\bar{Y}} = 2.60$; $z = (61 - 60)/2.60 = 0.38$. Prob. $= .3520$.

Problem 3. $\sigma_{\bar{Y}} = 6.32$. $Q_1 = 125 - .67 \times 6.32 = 120.77$, $Q_3 = 125 + .67 \times 6.32 = 129.23$.

Problem 4. $\sigma_{\bar{Y}} = 5.42$. $CI_{.95} = 75$ ($1.96 \times 5.42 = 75 \pm 10.62 = 64.38$ to 85.62.

Problem 5. Room B is more likely to be greater than 52.0. The smaller sample size makes it more likely of obtaining a residual greater than 2.

Problem 6. $s_{\bar{Y}} = 0.68$. $t(23) = 2.807$. Width of $CI_{.95} = 2 \times 2.807 \times 0.68 = 3.82$

Problem 7. The overall effect would be to reduce the variance and thus reduce the width of the 95% confidence interval.

Problem 8. None. They would all decrease the width of a confidence interval.

Problem 9. $\sigma_{\bar{Y}} = 3.0$. (a) .1587. (b) .905

Problem 10. (a) $t(22, .99) = 2.819$. The width of the confidence interval is 106 $= 2 \times s_{\bar{Y}} \times 2.819$. Thus $s_{\bar{Y}} = 106/(2 \times 2.819) = 18.80$. (b) s $= 18.80 \times \sqrt{23} =$

90.16. (c) $t(22, .95) = 2.074$. Note that the mean of the sample is the midpoint of the 99% confidence interval = 925. $CI_{.95} = 925 \pm 18.80 \times 2.074 = 925 \pm 39 = 886$ to 964.

Exercises in the Analysis of Realistic Data

Exercise 1
First sample: $s_{\bar{Y}} = 9.10/\sqrt{150} = 0.74$. $CI_{.95} = 45.7 \pm 0.74 \times 1.984 = 45.7 \pm 1.5 = 44.2$ to 47.2.

Second sample: $s_{\bar{Y}} = 9.67/\sqrt{65} = 1.20$. $CI_{.95} = 45.7 \pm 1.20 \times 2.000 = 48.8 \pm 2.4 = 46.4$ to 51.2. There are two sources of difference: 1. Smaller sample 2. Chance variation in the estimates of the mean and variance.

Exercise 2
Original study: $s_{\bar{Y}} = 28.43/\sqrt{200} = 2.01$. $CI_{.95} = 86.36 \pm 2.01 \times 1.96 = 86.36 \pm 3.94 = 82.4$ to 90.3. (Using the normal distribution approximation to the t-distribution)

Follow-up study: $s_{\bar{Y}} = 27.40/\sqrt{100} = 2.74$. $CI_{.95} = 88.06 \pm 2.74 \times 1.987 = 88.06 \pm 5.44 = 82.62$ to 93.50. (using $t(90, .95)$). Smaller sample results in larger standard error.

Exercise 3
Original study: $s_{\bar{Y}} = 14.22/\sqrt{120} = 1.30$. $CI_{.99} = 58.97 \pm 1.30 \times 2.626 = 58.97 \pm 3.41 = 55.56$ to 62.38. (using $t(100, .99) = 2.626$.)

Follow up study: $s_{\bar{Y}} = 17.49/\sqrt{120} = 1.60$. $CI_{.99} = 56.56 \pm 1.60 \times 2.626 = 56.56 \pm 4.20 = 52.36$ to 60.76. The difference in width results from the chance-determined higher value for the standard deviation (and thus of the standard error) of the second sample.

CHAPTER 6 Experiments with Two Independent Groups

Short-Answer Questions

1. a. T   b. F   c. F   d. F   e. T   f. T   g. F   h. F   i. T   j. F   k. T   l. F   m. T
    n. F   o. T   p. F   q. T   r. T   s. T   t. T
2. a. This study is neither a completely randomized design nor a quasi-experiment.
    b. This study is a quasi-experiment. c. This study is a completely randomized design.
3. See Section 6.2.3 in textbook.
4. The first engineer made the equivalent of a Type-2 (failing to detect error a genuine effect); the second engineer made the equivalent of a Type-1 error (a false alarm, claiming an effect when none exists.)

Short Problems

Problem 1. Critical $t(26) = 2.056$

Problem 2. Critical $t(34) = 2.728$

Problem 3. (a) We know that $t = 2.3 = 8.70/s_{\bar{Y}_1-\bar{Y}_2}$. Thus $s_{\bar{Y}_1-\bar{Y}_2} = 8.70/2.3 = 3.78$.

(b) $CI_{.99} = 8.70 \pm 2.712 \times 3.78 = 8.70 \pm 10.25 = -1.55$ to 18.95.

(c) No

(d) $Y = \mu + e$.

Problem 4. $w = 2.021 \times 4.4 = 8.89$.

Problem 5. a. (i) $7 - 10 = -3$ (model 1); $7 - 8 = -1$ (model 2)   (ii) $7 - 6 = 1$ (model 1); $7 - 8 = -1$ (control).

b. $SS_e = 45 + 55 = 100$. c. $MS_e = 100/58 = 1.724$. d. $\hat{d} = 4.0/\sqrt{1.724} = 3.05$

Problem 6. a. $s_{\bar{Y}_1-\bar{Y}_2} = \sqrt{\dfrac{2 \times 723}{15}} = 9.82$ b. $t = 2.048$ c. $w = 2.048 \times 9.82 = 20.11$.

d. $t = 12.8/9.82 = 1.30$ e. The null model: $Y = \mu + e$.

Exercises in the Analysis of Realistic Data

Exercise 1
a. $MS_e = 7.01$

b. $\hat{d} = 5/2.65 = 1.89$

c. $CI_{.95} = 5.0 \pm 2.021 \times 0.749 = 5.0 \pm 1.5 = 3.5$ to 6.5

d. $SS_{model} = 649.0 - 336.4 = 312.6$. Thus $R^2 = 312.6/649.0 = .48$.

e. The full model is justified. $t(48) = 5/0.749 = 6.68$ or $F(1,48) = 312.6/7.01 = 44.6$

f. The generate condition results in better remembering.

Exercise 2
a. $MS_e = 13.52$

b. $\hat{d} = 4.27/3.677 = 1.16$.

c. $CI_{.99} = 4.27 \pm 2.704 \times 1.109 = 4.27 \pm 3.00 = 1.27$ to 7.27.

d. $R^2 = SS_{model}/SS_{total} = 200.82/768.73 = .26$

e. The full model is justified.

f. Novices are poorer than experts in their memory for chess positions.

Exercise 3

a. $MS_e = 5.39$

b. $\hat{d} = 0.82/2.32 = 0.35$.

c. $CI_{.95} = 0.82 \pm 2.009 \times 0.621 = 0.82 \pm 1.25 = -0.43$ to 2.07

d. $R^2 = 9.45/300.55 = .03$

e. $t(54) = 0.82/0.621 = 1.32$.

f. There is no evidence for a difference between novices and experts in memory for randomly located chess positions.

Exercise 4

a. $MS_e = 5.76$

b. $\hat{d} = 2.87/2.4 = 1.20$

c. $CI_{.95} = 2.87 \pm 2.048 \times 0.877 \; 5 \; 2.87 \pm 1.80 = 1.07$ to 4.67

d. $R^2 = 61.63/222.97 = .28$.

e. $t(28) = 2.87/0.877 = 3.27$ Critical value is $t = 2.048$. Thus reject the null hypothesis and null model. Note that the confidence interval does not include 0, so this decision is consistent.

f. Training influences short-term memory.

Exercise 5

a. $MS_e = 3831.6/48 = 79.8$

b. $\hat{d} = 10.1/8.93 = 1.13$

c. $CI_{.95} = 10.1 \pm 2.021 \times 2.53 = 10.1 \pm 5.1 = 5.0$ to 15.2.

d. $R^2 = 1270.1/5101.7 = .25$

e. $t(48) = 10.1/2.53 = 3.99$. Critical value is $t = 2.021$ (using $df = 40$). Thus reject the null hypothesis and null model. Note that the confidence interval does not include 0, so decision is consistent.

f. Instructions from parents and strangers result in differences in speed of compliance; it appears that children comply more readily to the request from a stranger.

Exercise 6

a. $MS_e = 26.55$

b. $\hat{d} = 4.9/5.15 = .95$

c. $CI_{.95} = 4.9 \pm 2.021 \times 1.52 = 4.9 \pm 3.07 = 1.8$ to 8.0.

   d. $R^2 = 272.7/1440.9 = .19$.

   e. $t(44) = 4.87/1.52 = 3.20$ Critical value is $t = 2.021$ (using $df = 40$). Thus reject the null hypothesis and null model. Note that the confidence interval does not include 0, so decision is consistent.

   f. Mood influences speed of compliance; it appears children in a good mood are more ready to comply.

Exercise 7

   a. $MS_e = 51.32$

   b. $\hat{d} = 6.1/7.16 = .85$

   c. $CI_{.95} = 6.1 \pm 2.032 \times 2.39 = 6.1 \pm 4.9 = 1.2$ to 11.0.

   d. $R^2 = 336.11/2080.89 = .16$.

   e. $t(34) = 6.1/2.39 = 2.55$. Critical value is $t = 2.032$. Thus reject the null hypothesis and null model. Note that the confidence interval does not include 0, so decision is consistent.

   f. Attention influences speed of compliance; it appears that it is better to obtain the child's attention before issuing a request.

CHAPTER 7 Larger Experiments with Independent Groups: Analysis of Variance

Short-Answer Questions

1. Simultaneous confidence intervals will be wider than individual confidence intervals calculated on the same data. See Section 7.1.4 in textbook.

2. An experiment with four conditions has six different comparisons. The confidence intervals for all these comparisons will have the same width because they are based on a single common estimate of the standard error. The assumption of homogeneity of variance.

3. Greater protection is needed against the possibility of any one interval not including the true value.

4. See Section 7.1.4 in textbook.

5. The width of simultaneous confidence intervals is twice Tukey's HSD.

6. When the F-ratio is significant.

7. SS_{model} and $SS_{between}$ are the same. $SS_{total} = SS_{between} + SS_{within}$.

8. See Section 7.3.1 in textbook.

9. $SS_{between}$ is the the sum of these three sums of squares.

10. Line drawn joining the cell means should be parallel.

Short Problems

Problem 1 (a) $df = 30$, $k = 5$; $q = 4.10$. (b) $t(30) = 2.042$. (c) Fisher's LSD is a per-comparison value whereas Tukey's HSD is a per-experiment value.

Problem 2. (a) $df = 60$, $k = 4$; $q = 3.74$. (b) $MS_e = 210/60 = 3.5$. Tukey's HSD ($\alpha = .05$) $= 3.74 \times 0.47 = 1.76$. For conditions A_1 and A_2 $CI_{.95} = 4.3 \pm 1.76$; for conditions A_1 and A_4 $CI_{.95} = 3.3 \pm 1.76$. (c) All but A_2 versus A_4. (d) Width $= 2 \times 1.76 = 3.52$.

Problem 3

$SS_{between} = n(a_1^2 + a_2^2 + a_3^2) = 12(2^2 + 3^2 + 1^2) = 168.0$

Source	SS	df	MS
Between conditions	168	2	84
Within conditions	495	33	15
Total	663	35	

$F(2,33) = 84/15 = 5.6$, which exceeds the critical tabled value of F. Thus the results justify the use of Fisher's LSD to test comparisons. LSD $= t(33) \times \sqrt{\dfrac{2MS_e}{n}} = 2.035 \times 1.58 = 3.22$.

Problem 4 (a) $MS_{between} = (195.0 - 162.0)/4 = 8.25$ (b) $MS_{within} = 162.0/70 = 2.31$ (c) The critical F-ratio with 4 and 70 degrees of freedom $= 2.50$.

Problem 5

Source	SS	df	MS
Between conditions	48	3	16
Within conditions	144	36	4
Total	192	39	

The critical F-ratio with 3 and 36 degrees of freedom $= 2.87$. The obtained F-ratio is $16/4 = 4$ which is greater than the critical value. Thus reject the null hypothesis and accept the full model.

Problem 6 a. $MS_{between} = 3.18$ b. $MS_{within} = 1.16$ c. $F = 2.75$.

Problem 7 a. $Y_{ij} = \mu + \alpha_i + \beta_j + \alpha\beta_{ij} + e$ b. The interaction term $\alpha\beta_{ij}$.   c. F with 1 and 32 degrees of freedom $= 4.15$.

d.

Source	SS	df	MS
Drug (D)	11	1	11
Motivation (M)	25	1	25
D × M	33	1	33
Within conditions	145	32	4.53

e.   M and D × M.

f.   Yes

Problem 8

(a) $F = 3.22$ (b) $F = 4.07$ (c) $F = 3.22$

Source	SS	df	MS	F
Model	32.7	5		
Frustration (F)	16.8	2	8.4	4.24
Delay (D)	9.8	1	9.8	4.95
F × D	6.1	2	3.05	1.54
Within conditions (residual)	83.1	42	1.98	
Total		47		

The additive model: $Y_{ij} = \mu + \alpha_i + \beta_j + e$

Problem 9

	$F_{obtained}$	$F_{critical}$
Fatigue (A)	2.89	3.14
Difficulty (B)	5.17	3.99
A×B	4.59	3.14

Problem 10
Experiment 1: B, A×B; Experiment 2: A; Experiment 3: B, A×B; Experiment 4: A, B

Exercises in the Analysis of Realistic Data

Exercise 1
1. SS_e = 4083.8; MS_e = 80.07. The error bars indicating the standard error of the means should extend 2.1 above and below each bar.

2. Tukey's HSD = 3.44 × 2.11 = 7.26. For $\mu_1 - \mu_2$ CI$_{.95}$ = 11.61 ± 7.26; For $\mu_1 - \mu_3$ CI$_{.95}$ = 26.66 = 7.26; For $\mu_2 - \mu_3$ CI$_{.95}$ = 15.05 ± 7.26.

3. Yes. The results suggest that the presence of the cue word provides some facilitation, but not as much as in the generate condition.

4. R^2 = .61.

5.

Source	SS	df	MS	
Between conditions	6,435.59	2	3217.8	F = 40.2
Within conditions	4,083.83	51	80.07	
Total	10,519.42	53		

The critical value of the F-ratio is 3.18. Therefore reject the null hypothesis.

The null hypothesis that this F-test evaluates is $H_0 = \mu_1 = \mu_2 = \mu_3$.

Exercise 2
1. SS_e = 3841.0; MS_e = 75.31. The error bars indicating the standard error of the means should extend 2.05 above and below each bar.

2. Tukey's HSD = 3.44 × 2.05 = 7.05. For $\mu_1 - \mu_2$ CI$_{.95}$ = −14.50 ± 7.05; for $\mu_1 - \mu_3$ CI$_{.95}$ = −23.89 ± 7.05; for $\mu_2 - \mu_3$ CI$_{.95}$ = −9.39 ± 7.05. (Note: Rather than using absolute differences, the direction of each difference has been retained to maintain comparability with Exercise 1.)

Source	SS	df	MS
Between conditions	5,214.5	2	2,607.24
Within conditions	3,841.0	51	75.31
Total	9,055.5	53	

3. R^2 = .58.

4. The results indicate strong differences between implicit and explicit memory with respect to the generation effect. The ordering of the conditions reverses. Plotting the results from both experiments on a single graph makes this fact very clear.

Exercise 3

1. $SS_e = 1100.32$; $MS_e = 15.3$. The error bars indicating the standard error of the means should extend 0.78 above and below each bar.

2.

Source	SS	df	MS	
Between conditions	792.83	2	396.4	$F = 25.9$
Within conditions	1,100.32	72	15.3	
Total	1,893.15	74		

$F(2,72) = 25.9$ is greater than the tabled value for $\alpha = .05$ and so Fisher's LSD can be applied. LSD $= t(72) \times s_{\bar{Y}_1 - \bar{Y}_2} = 1.994 \times 1.11 \times 1.06 = 2.21$. All three comparisons exceed LSD.

3. $R^2 = .42$.

Exercise 4

1. $SS_e = .7028$; $MS_e = .0117$. The error bars indicating the standard error of the means should extend 0.02 above and below each bar.

2.

Source	SS	df	MS	
Between conditions	.040	2	.020	$F = 1.71$
Within cconditions	.70	60	.0117	
Total	.743	62		

$F(2,60) = 1.71$ is less than the tabled value for $\alpha = .05$ ($F = 3.15$) and so Fisher's LSD cannot be applied.

3. All sums of squares and mean squares would be multiplied by 40^2. Because the F-ratios are one mean square divided by another, this multiplying factor of 40^2 cancels out, leaving the F-ratio unchanged.

4. $R^2 = .05$.

5. No.

Exercise 5

1. Full model: $Y_{ij} = \mu + \alpha_i + \beta_j + \alpha\beta_{ij} + e$ Null model: $Y_{ij} = \mu + e$

2.

Source	SS	df	MS	F-ratio
Model: Between conditions	1,905.0	5		
A (Level)	1,876.1	2	938.1	63.3
B (Instruction)	17.8	1	17.8	1.2
A × B	11.1	2	5.5	< 1
Residuals: Within conditions	1,245.2	84	14.8	
Total	3,150.2	89		

3. $R^2 = .60$

4. There is no evidence that incidental versus intentional instructions have an effect on the levels of processing.

Exercise 6

1. $SS_e = 66.53$; $MS_e = 1.58$. The error bars indicating the standard error of the means should extend 0.32 above and below each bar.

2. Tukey's HSD = 3.44 × 0.32 = 1.1. For $|\mu_1 - \mu_2|$ $CI_{.95} = 1.3 \pm 1.1$. For $|\mu_1 - \mu_3|$ $CI_{.95} = 1.4 \pm 1.1$; For $|\mu_2 - \mu_3|$ $CI_{.95} = 0.1 \pm 1.1$.

Source	SS	df	MS
Between conditions	18.71	2	9.36
Within conditions	66.53	42	1.58
Total	85.24	44	

3. $R^2 = .22$.

4. Condition 1 differs from both conditions 2 and 3 which themselves do not differ.

Exercise 7

1. The error bars indicating the standard error of the means should extend 2.8 above and below each bar.

2. Tukey's HSD = 3.69 × 2.84 = 10.5. No differences are significant.

Exercise 8

1. It appears as if there could be an interaction between gender and school in that for schools B2 and B4 males perform more poorly than females. Does formal analysis support this conjecture?

2. Full model: $Y_{ij} = \mu + \alpha_i + \beta_j + \alpha\beta_{ij} + e$ Null model: $Y_{ij} = \mu + e$

3.

Source	SS	df	MS	F-ratio
Model: Between conditions	3,858.1	7		
A (Gender)	2,070.3	1	2070.3	2.96
B (School)	6,22.1	3	207.4	< 1
A × B	11,65.7	3	388.6	< 1
Residuals: Within conditions	27,4354.3	392	699.9	
Total				

No effects are significant.

4. $R^2 = .014$

5. There is no evidence for sex differences in scholastic aptitude among the four schools?

CHAPTER 8 Increasing the Precision of an Experiment

Short-Answer Questions

1. Power $(1 - \beta)$ or $2w$, the width of the confidence interval.

2. Reduce MS_e, increase the sample size, n.

3. $1 - \beta$.

4. Score on the Math section of the SAT is more likely to result in the greater increase in precision because it is likely to be a stronger determinant of individual differences in algebra performance within each condition.

5. See Section 8.2.5 in textbook.

6. Repeated measures can be thought of as a special case of matching.

7. $SS_{model} = SS_{between} + SS_{subjects}$

8. This result suggests that the variable on which the blocking was based is unrelated to the response measure and thus does not account for any of the variance within conditions.

Short Problems

Problem 1. $n = 2 \times 150^2 \times (1.96/50)^2 = 70$.

Problem 2. $d = 1.0/1.5 = 0.67$. $1 - \beta = .7$, Using $d = 0.65$, $n = 30$.

Problem 3. One way would be to ensure an adequate time interval between conditions. $d = 1.0/1.1 = 0.90$. $n = 8$.

Problem 4. $n = 2 \times 92 \times (1.96/5)^2 = 25$. (a) .5 (b) $d = 3/9 = 0.33$, $1 - \beta =$ is approximately .2.

Problem 5. a. With $df = 31$, $t = 2.040$. b. Width $= 2 \times 2.04 \times 2.3 = 9.38$. c. 4.69. d. Note that $s_D = 2.3 \times \sqrt{32} = 13.0$, so $d = 4.0/13.0 = 0.31$. From the table for matched-pair designs, $n = 69$.

Problem 6. a. $CI_{95} = 4.8 \pm 2.037 \times 2.26 = 4.8 \pm 4.6$ b. $CI_{95} = 3.2 \pm 2.447 \times 1.81 = 3.2 \pm 4.4$ c. $CI_{95} = 4.1 \pm 2.023 \times 1.51 = 4.1 \pm 3.1$

Problem 7. a. Full b. Null c. Null

Problem 8. $SS_{between} = 15[0.1^2 + 0^2 + (-0.1)^2]$

Source	SS	df	MS
Model	5.260	15	
Between Conditions	0.300	2	.15
Participants	4.960	14	.354
Residuals	1.014	28	.036
Total	6.274	44	

$F(2, 28) = 0.15/0.036 = 4.17$. For $\alpha = .05$, the critical value is 3.34. Therefore reject the null hypothesis. HSD $= 3.52 \times 0.05 = 0.17$. For $|\mu_1 - \mu_2|$ $CI_{.95} = .1 \pm 0.17$; For $|\mu_1 - \mu_3|$ $CI_{.95} = .2 \pm 0.17$; For $|\mu_2 - \mu_3|$ $CI_{.95} = .1 \pm 0.17$.

Exercises in the Analysis of Realistic Data

Exercise 1. From Exercise 3 of Chapter 6 we have $MS_e = 5.39$ and $CI_{.95} = -0.43$ to 2.07, a width of 2.5. That is, the experiment has a probability of $1- \beta = 1.5$ of detecting a difference of 1.25. A difference of 1.25 corresponds to an effect size of $d = 1.25/\sqrt{5.39} = .54$. Thus the experiment has only a 50% chance of detecting even a moderate effect.

Exercise 2

Source	SS	df	MS
Model	360.30	25	
Between conditions	89.78	1	89.78
Participants	270.52	24	11.27
Residuals	31.72	24	1.32
Total	392.02	49	

a. $s_{\bar{D}} = \sqrt{2\times1.32/25} = 0.325$

b. $s_D = .325\sqrt{25} = 1.625$. Thus $\hat{d} = 2.68/1.625 = 1.65$

c. $CI_{.95} = 2.68 \pm 2.064 \times .325 = 2.68 \pm 0.67 = 2.01$ to 3.35

d. $R^2 = 360.30/392.02 = .92$

e. The data justify the full model.

f. The results support the claim that the "generation" condition produces better remembering than the "read" condition.

g. MS_e is reduced from 7.01 to 1.32. R^2 is increased from .48 to .92.

Exercise 3

a. $Y_{ij} = \mu + \alpha_i + \pi_j + + e$

b.

Source	SS	df	MS	
Model	10,034.0	19		
Between conditions	5,809.3	2	2,904.7	$F = 69.7$
Participants	4,224.7	17	248.5	
Residuals	14,16.0	34	41.65	
Total	11,450.0	53		

c. HSD $= 3.49 \times 1.52 = 5.3$ For $|\mu_1 - \mu_2|$ $CI_{.95} = 14.3 \pm 5.3$; for $|\mu_1 - \mu_3|$ $CI_{.95} = 25.3 \pm 5.3$; for $|\mu_3 - \mu_4|$ $CI_{.95} = 11.00 \pm 5.3$.

d. Yes

e. $R^2 = 10034/11450 = .88$

f. MS_e is reduced from 80.08 to 41.65. R^2 is increased from .61 to .88.

g. There are possible carryover effects. Any cross-condition influence on the way a word is studied would constitute a carryover effect.

Exercise 4

a.

Source	SS	df	MS
Model	9,229.3	17	
Between conditions	5,226.0	2	2613.0
Participants	4,003.3	15	266.9
Residuals		30	44.3
Total	10,559.3	47	

b. $F(2,30) = 59.0$. Reject H_0.

c. HSD $= 3.49 \times 1.66 = 5.79$ For $|\mu_1 - \mu_2|$ $CI_{.95} = 14.25 \pm 5.79$; For $|\mu_1 - \mu_3|$ $CI_{.95} = 25.5 \pm 5.79$; For $|\mu_2 - \mu_3|$ $CI_{.95} = 11.25 \pm 5.79$.

d. Yes

e. $R^2 = 9229.3/10559.3 = .87$

f. Because the measurement scales are different a direct comparison of MS_e is not meaningful. Note, however, that R^2 has increased from .42 to .87. A further comparison could be made by calculating and comparing values of $1 - \beta$ for a specified value of **d**.

g. There are possible carryover effects. Any cross-condition influence on the way a word is studied would constitute a carryover effect.

Exercise 5

a. $n = 2 \times 26.55 \times (1.96/3)^2 = 23$.

b. $\hat{d} = 4/5.15 \times 0.78$. This value gives a power of $1 - \beta$ of approximately .7

Exercise 6

a.

Source	SS	df	MS
Model	70.89	16	
Between conditions	24.58	2	12.29
Participants	46.31	14	3.31
Residuals	22.76	28	0.81
Total	93.65	44	

b. HSD $= 3.52 \times 0.23 = 0.81$ For $|\mu_1 - \mu_2|$ $CI_{.95} = 1.6 \pm 0.81$; For $|\mu_1 - \mu_3|$ $CI_{.95} = 1.53 \pm 0.81$; For $|\mu_2 - \mu_3|$ $CI_{.95} = 0.07 \pm 0.81$.

c. Yes

d. $R^2 = 70.89/93.65 = .76$

e. MS_e is reduced from 1.58 to 0.81. R^2 is increased from .22 to .76.

f. There are possible carryover effects. A parent may be influenced by past judgments in the experiment to take a perspective other than the one instructed.

CHAPTER 9 Quantitative Predictor Variables:
Linear Regression and Correlation

Short-Answer Questions

1. a. conditional distribution b. marginal distribution c. marginal distribution
   d. conditional distribution

2. The conditional distribution; it would have a smaller variance.

3. b. greater than .37

4. False. Such restriction of range will reduce the correlation.

5. These numbers would result in a correlation coefficient greater than 1.

6. a. No effect.

7. False; there will be regression toward the mean.

8. Variance is an average sum of squares and covariance is an average sum of products.

9. Homogeneity of variance and normality of the conditional distributions; linear regression.

10. Restricted range

Short Problems

Problem 1 a. Slope = .563   b. Predicted GPA = 3.16

Problem 2

Source	df
Model: Linear regression	1
Residual	38
Total	39

Problem 3. (a) Residuals could be plotted, or a graph of the four conditional means

b.

Source	df
Model:	3
Linear regression	1
Departure from linear regression	2
Residual:	36
Total	39

Problem 4. a. Time drops 30 minutes in 15 hours of training. The slope $= -30/15 = -2.0$. The intercept (0 hours) would be 40+10 or 50 hours.

b. Time after 15 hours of training $= 50 - 15 \times 2.0 = 20$ minutes.

Problem 5.   a. 3736.0./149 = 25.07   b. r = 25.07/(5.3 × 9.4) = .503.
c. b = 0.284   d. b = 0.89 e. The standard deviations are different.

Problem 6

Source	SS	df	MS
Model: Linear regression	41.3	1	41.3
Residual	205.0	119	1.72
Total	246.3	39	

a. $R^2 = 41.3/246.3 = .168$.   b. Standard error of estimate $= \sqrt{1.72} = 1.31$
c. $F(1,119) = 24.0$ which exceeds the critical value. There is a significant linear relationship.

Exercises in the Analysis of Realistic Data

Exercise 1. The plot of the means suggests that a linear regression model is implausible; forgetting is curvilinear.

Exercise 2

a.

Source	SS	df	MS
Model: Linear regression	612.51	1	612.51
Residual	2,862.48	63	45.44
Total	3,474.99	64	

b. $R^2 = 612.51/2862.48 = .176$. $r = \sqrt{.176} = -.42$.

Alternatively, using the covariance, $r = -29.90/(9.67 \times 7.40) = -.42$.

c. $b = -0.32$, $s_b = 0.087$.

$CI_{.95} = -0.320 \pm 2.000 \times 0.087 5 -0.320 \pm 0.174 = -0.494$ to -0.146.

d. The standard error of estimate $= \sqrt{45.44} = 6.74$.

e. Predicted grade $= 83.8 - 0.32 \times$ mother's authoritarianism score

f. High scores on authoritarianism of the mother are associated with lower school grades.

Exercise 3
a. $R^2 = SS_{regression}/SS_{total} = 841.76/3634.74 = .232$. $r = \sqrt{.232} = .48$.

Alternatively, using the covariance, $r = 33.44/(8.85 \times 7.85) = .48$.

b. $b = 0.427$, $s_b = 0.102$.

$CI_{.95} = 0.427 \pm 2.009 \times 0.102 = 0.427 \pm 0.205 = 0.22$ to 0.63.

c. Predicted concern $= 9.41 + 0.427 \times$ chores

d. High scores on routine chores are associated with higher levels of concern for others.

Exercise 4

a. $R^2 = SS_{regression}/SS_{total} = 5.37/1323.52 = .004$. $r = .06$.

b. $b = 0.042$, $s_b = 0.095$.

   $CI_{.95} = 0.042 \pm 2.021 \times 0.095 = 0.042 \pm 0.192 = -0.15$ to 0.23.

c. There is no evidence that concern for others related to the performance of ad hoc household chores.

Exercise 5

a. Stemplots for each administration would indicate any problematic features of the marginal distribution. A scatterplot would indicate any obvious abnormality in the bivariate distribution.

b. $r = 66{,}817.13/(114 \times 25.94 \times 26.77) = .844$

Exercise 6

a. See answer to (a) in Exercise 5.

b.

Source	SS	df	MS	
Model: Linear regression	5,434.9	1	5,434.9	F = 116.5
Residual	5,036.9	108	46.6	
Total	10,471.8	109		

c. $a = 48.0$, $b = 0.248$. Thus predicted grade $= 48.0 + 0.248 \times$ aptitude test score.

d. (i) The standard error of estimate $= 6.83$ (ii) $s_b = .023$; using $t = 1.984$, $CI_{.95} = 0.202$ to 0.294.

e. $r = .720$

Exercise 7

a.

Source	SS	df	MS	
Model: Linear regression	1,907.5	1	1,907.5	F = 19.15
Residual	7,271.2	73	7,271.2	
Total	9,178.7			

b. $r = .456$

c. $b = 0.306$, $s_b\ 0.070$, $CI_{.95} = 0.167$ to 0.445.

CHAPTER 10 Categorical Response Variables and Distribution-Free Methods

Short-Answer Questions

1. χ^2 with $df = 4$ is 9.49

2. χ^2 with $df = 6$ is 12.59

3. a. T   b. F   c. T   d. F   e. F

4. f_e.

5. See Section 10.1.2 in textbook.

6. Small expected frequencies.

Short Problems

Problem 1. Proportion of unemployed women who win this lottery = .65 $\times$ .15 = .0975. It would mean that there was the same proportion of women winning regardless of employment status.

Problem 2. The three expected frequencies are each 96/3 = 32. Thus the residuals are -2, -14, and 16. χ^2 with $df = 2$ is therefore 14.25. This exceeds the critical value (for $\alpha = .05$, or .01), thus these data are strong enough to reject the hypothesis that the consumers show no brand preference.

Problem 3. The 2 $\times$ 3 contingency table gives $\chi^2(2) = 9.86$. This exceeds the critical value (for $\alpha = .05$, or .01), thus these data support the hypothesis of sex differences in the pattern of brand preference.

Problem 4

	Day	Eve	
A	50	100	$\chi^2(1) = 11.1$
B	30	20	

Problem 5. The four expected frequencies are each 100/4 = 25. Thus the residuals are -4, 6, 4, and -6. χ^2 with $df = 3$ is therefore 4.16. This is less than the critical value (for $\alpha = .05$), thus these data these results support the claim that the machine is shuffling properly.

Problem 6

	Stopped	Not stopped	
Experimental	23	13	$\chi^2(1) = 6.727$
Control	12	24	

The data support the claim that the experimental procedure was effective.

Problem 7

	Favor	Opposed
Republican	37	28
Democrat	48	27

$\chi^2(1) = 0.73$

The contingency is not statistically significant.

Problem 8

	Yes	No
Men	14	6
Women	7	13

$\chi^2(1) = 4.912$

The contingency is statistically significant.

Problem 9

	Normal	High
Depressive	10	30
Non-depressive	23	21

$\phi = .279$

$\chi^2(1) = 6.534$

Exercises in the Analysis of Realistic Data

Exercise 1

	< 21	21–34	35–44	>44	Total
f_o	26	95	44	35	200
f_e	14	60	44	82	200

$\chi^2(3) = 57.6$.

Exercise 2

a. The critical value of $\chi^2(1) = 3.84$

CHANGE IN AGGRESSION

	Increase	No increase
Neutral movie	11	35
Violent movie	31	15

b. $\chi^2(1) = 17.5$ therefore reject H_0.

c. $\phi = .436$

Exercise 3

	Agreed	Disagreed
Group A	21	9
Group B	12	18

$\phi = .302$

$\chi^2(1) = 5.455$

The critical value of $\chi^2(1) = $ is 3.84, therefore reject H_0.

Exercise 4

	S	IA	IR
Mother at home	40	5	11
Mother at work	29	17	8

$\chi^2(1) = 8.739$. For $\alpha = .05$, the critical value of $\chi^2(2) = $ is 5.99, therefore reject H_0.

The two categorical variables used in this study are natural. Care should therefore be taken in making causal interpretations of the data. The data indicate that mothers at home have a higher proportion of secure infants than do mothers at work, but it does not follow that being or not being at home is the cause of this effect.

Appendix B
Data Sets Used in Textbook and Study Guide

All of the data sets used in the textbook and the study guide are available as text (ASCII) files. You can either get them from your instructor or down load them from the book's web site. It should prove easy to locate any specific data set you wish, once you understand our naming convention.

Data sets from the textbook have filenames corresponding to their designations in the textbook itself. Thus Data Set 2.1 in the textbook adopts the filename **DS02_01.dat**; Data Set 2.2 has the filename **DS02_02.dat**; Data Set 3.1 has the filename **DS03_01.dat**, and so forth. The leading zeros (01, 02, etc.) have been inserted so that when the files are listed on your screen they will be correctly ordered. Without the leading zeros, most software when reading filenames will place the number 11 before the number 2. However, 02 will be placed before 11. The filename extension ".dat" has been added because both SPSS and SAS look for this extension as designating a data set.

Data sets from the study guide have filenames denoting one of the three research areas used in the study guideexercises. Data sets for the memory area are named **MEM_01.dat**, **MEM_02.dat**, etc. Data sets for the child behavior area are named **CHILD_1.dat**, **CHILD_2.dat**, etc. Data sets for the exercises involving test construction and evaluation are named **TEST_1.dat**, **TEST_2.dat**, etc.

For convenient reference, all data sets are listed in the following table along with a descriptive label. The column headed "variable names" provides suggested labels (eight letters maximum) that can be used when reading in the data set to either the SPSS or the SAS programs. Both programs require that you enter a label for each variable, and although you are free to use your own label, the suggested names have been chosen to provide a meaningful reference to the particular data of that exercise.

Data Sets Used in Textbook

Textbook designation	ASCII filename	Suggested variable names	Description
Data Set 2.1	DS02_01.dat	response	Infant imitative gestures
Data Set 2.2	DS02_02.dat	choice	Brand choice
Data Set 2.3	DS02_03.dat	IQ	IQ scores
Data Set 2.4	DS02_04.dat	group [1=6 years old, 2=8 years old]; vocab	Age and vocabulary
Data Set 2.5	DS02_05.dat	expressn	Facial expressions
Data Set 2.6	DS02_06.dat	group [1=same sex, 2=different sex]; interact	Toddler interactions
Data Set 2.7	DS02_07.dat	grade	Final grades
Data Set 2.8	DS02_08.dat	Y1;Y2;Y3;Y4;Y5; Y6;Y7;Y8;Y9;Y10	Section 2.2 data for Problem 5
Data Set 2.9	DS02_09.dat	Y1;Y2;Y3;Y4;Y5; Y6;Y7;Y8	Section 2.2 data for Problem 8
Data Set 2.10	DS02_10.dat	Y1;Y2;Y3;Y4;Y5	Section 2.2 data for Problem 9
Data Set 2.11	DS02_11.dat	group [1=one-visit; 2=two-visit]; response	Foot-in-the-door phenomenon
Data Set 2.12	DS02_12.dat	payment [1=$1; 2=$100]; rating	Opinion change
Data Set 2.13	DS02_13.dat	reward [1=reward; 2=no reward] time	Undermining intrinsic motivation
Data Set 2.14	DS02_14.dat	time	Intermodal matching
Data Set 2.15	DS02_15.dat	time	Face recognition
Data Set 2.16	DS02_16.dat	stimulus [1=non-occluded, 2=aligned]; time	Infant perception of partly occluded stationary object
Data Set 2.17	DS02_17.dat	Atwin; Btwin	IQ scores for identical twins
Data Set 2.18	DS02_18.dat	verbal; perform	WAIS verbal and performance scores
Data Set 3.1	DS03_01.dat	condit [1=control, 2=desensitization]; heartrt	Phobia treatment experiment
Data Set 3.2	DS03_02.dat	threat [1=mild, 2=severe]; rating	Effect of threat on attitude
Data Set 5.1	DS05_01.dat	estimate	Ponzo illusion
Data Set 5.2	DS05_02.dat	time	Role of smell in infant perception
Data Set 6.1	DS06_01.dat	condit [1=control, 2=experimental]; time	Tranquilizer effects

Textbook designation	ASCII filename	Suggested variable names	Description
Data Set 6.2	DS06_02.dat	condit [1=control, 2=experimental]; haprate	Induced happiness
Data Set 6.3	DS06_03.dat	condit [1=control, 2=experimental]; angerate	Induced anger
Data Set 6.4	DS06_04.dat	condit [1=incidental, 2=intentional]; recall	Incidental and intentional remembering
Data Set 6.5	DS06_05.dat	stimulus [1=non-occluded, 2=aligned]; time	Infant perception of partly moving stationary object
Data Set 6.6	DS06_06.dat	intstruc [1=normal, 2=imagery]; recall	Memory and mental imagery
Data Set 6.7	DS06_07.dat	intstruc [1=normal, 2=mnemonic]; recall	Method of locic
Data Set 7.1	DS07_01.dat	therapy [1=control 2=syst. desence., 3=countercondit.]; heartrt	Phobia treatment
Data Set 7.2	DS07_02.dat	tranq [1=control, 2=T1,3=T2]; time	Tranquilizer effects
Data Set 7.3	DS07_03.dat	emotion [1=anger, 2=sadness, 3=happiness]; heartrt	Induced emotion
Data Set 7.4	DS07_04.dat	instruct [1=intentional, 2=incidental, 3=both]; recall	Intentional and incidental remembering
Data Set 7.5	DS07_05.dat	agegroup [1=A, 2=B,3=C]; vocab	Age and vocabulary
ASCII file	DS07_06.dat	agegroup [1=A, 2=B,3=C]; recall	Age and memory
Data Set 7.7	DS07_07.dat	alcohol [1=none, 2=1 ounce]; sleepdep [1=none, 2=24 hours]; time	Alcohol and sleep deprivation
ASCII file	DS07_08.dat	therapy [1=control 2=syst. desence., 3=countercondit.]; time [1=short, 2=long]; rating	Therapy, duration of therapy, and anxiety
ASCII file	DS07_09.dat	condit [1=control, 2=tranquilizer]; sex [1=female, 2=male]; time	Tranquilizer effects
Data Set 7.10	DS07_10.dat	acquisit [1=sober, 2=intox]; retriev [1=sober, 2=intox]; recall	State dependent retrieval

Textbook designation	ASCII filename	Suggested variable names	Description
ASCII file	DS07_11.dat	agegroup [1=A, 2=B,3=C]; rate [1=fast, 2=slow]; recall	Age, memory, and rate of presentation
ASCII file	DS07_12.dat	condit [1=control, 2=$1, 3=$100]; rating	Extension of opinion change study
ASCII file	DS07_13.dat	reward [1=reward expected, 2=no reward, 3=surprise reward]; time	Extension of experiment on undermining intrinsic motivation
ASCII file	DS07_14.dat	reward [1=reward expected, 2= no reward, 3=surprise reward]; sex [1=girl, 2=boy], time	Further extension of experiment on undermining intrinsic motivation
Data Set 8.1	DS08_01.dat	P; exptal; control	Matched-pair version of Jill's experiment
Data Set 8.2	DS08_02.dat	P; happy; neutral	Induced happiness
Data Set 8.2	DS08_02a.dat	P; condit [1=happy 2=neutral]	Induced happiness (same data in different file format)
Data Set 8.3	DS08_03.dat	P; exptal; control	Tranquilizer effects
Data Set 8.3	DS08_03a.dat	P; condit [1=exptal, 2=control]; time	Tranquilizer effects (same data in different file format)
Data Set 8.4	DS08_04.dat	P; rate; count	Incidental remembering
Data Set 8.5	DS08_05.dat	P; exptal; control	Tranquilizer effects
Data Set 8.5	DS08_05a.dat	P; condit [1=exptal, 2=control]	Tranquilizer effects (same data in different file format)
Data Set 8.6	DS08_06.dat	P; before; after	Evaluation of typing speed
Data Set 8.7	DS08_07.dat	P; condit [1=pretest, 2=immediate posttest, 3=delayed posttest]; score	Attribution and math performance
Data Set 8.8	DS08_08.dat	P; delay [1=1 week, 2=3 months, 3=1 year]; anxiety	Therapy follow-up
Data Set 8.9	DS08_09.dat	P; condit [1=control, 2=imagery, 3=tutorial]; score	Improving geometry teaching

Textbook designation	ASCII filename	Suggested variable names	Description
Data Set 8.10	DS08_10.dat	P; delay [1=1 week, 2=3 months, 3=1 year, 4=2 years]; rating	Therapy follow-up
Data Set 8.11	DS08_11.dat	P; level [1=appearance, 2=sound, 3=meaning];	Levels of processing
Data Set 8.12	DS08_12.dat	P; color [1=red, 2=orange, 3=yellow, 4=green, 5=blue]; peckrate	Color stimulus generalization in pigeons
Data Set 8.13	DS08_13.dat	P; mother; stranger	Photograph recognition
Data Set 8.14	DS08_14.dat	P; habit; novel	Perception of facial expressions
Data Set 8.15	DS08_15.dat	P; stimulus [1=aligned, 2=nonoccluded]; time	Infant perception of partly occluded stationary object.
Data Set 8.16	DS08_16.dat	P; stimulus [1=aligned, 2= nonoccluded]; time	Infant perception of partly occluded moving object
Data Set 8.17	DS08_17.dat	P; stimulus [1=face, 2=scrambled, 3=black]; rotation deplevel; errors	Perception of scrambled faces
Data Set 9.1	DS09_01.dat		Sleep deprivation and concentration
Data Set 9.2	DS09_02.dat	age; score	Test scores and age
Data Set 9.3	DS09_03.dat	IQ; GPA	IQ and GPA
Data Set 9.4	DS09_04.dat	alcohol; time	Alcohol consumption and motor coordination
Data Set 9.5	DS09_05.dat	verbal; numerical	Verbal comprehension and numerical ability
Data Set 9.6	DS09_06.dat	THClevel; recall	Memory and marijuana
Data Set 9.7	DS09_07.dat	stutime; recall	Total time hypothesis
Data Set 9.8	DS09_08.dat	income; hapness	Happiness and income
Data Set 9.9	DS09_09.dat	rating; time	Maternal attention and infant attractiveness
Data Set 9.10	DS09_10.dat	dosage; depressn	Depression and drug dosage level
Data Set 9.11	DS09_11.dat	anxiety; score	Anxiety and performance

Textbook designation	ASCII filename	Suggested variable names	Description
Data Set 9.12	DS09_12.dat	practice; recall	Recall practice and recall
Data Set 9.13	DS09_13.dat	Atwin; Btwin	IQs of fraternal twins reared together
Data Set 9.14	DS09_14.dat	Atwin; Btwin	IQs of identical twins reared apart
Data Set 9.15	DS09_15.dat	Atwin; Btwin	IQs of identical twins reared together
Data Set 9.16	DS09_16.dat	verbal; perform	WAIS subscales, random sample
Data Set 9.17	DS09_17.dat	verbal; perform	WAIS subscales, restricted sample
Data Set 9.18	DS09_18.dat	group [1= below average,2=gifted] verbal; perform	WAIS subscales for a combined sample
Data Set 10.1	DS10_01.dat	condit [1=one-visit, 2=two-visit initial request, 3=two-visit familiarization only]; response	Familiarization and foot-in-the-door phenomenon
Data Set 11.1	DS11_01.dat	group [1=natural science, 2=humanities, 3=social sciences, 4=psychology]; score	Statistical reasoning in different college groups
Data Set 11.2	DS11_02.dat	P; preECT, postECT	Evaluation of ECT
Data Set 11.3	DS11_03.dat	P; condit [1=pre ECT, 2=3 ECT, 3=6 ECT]; depressn	Extended evaluation of ECT
Data Set 11.4	DS11_04.dat	group [1=no reward, 2=delay reward;3=reward]; errors	Latent learning at day 6
Data Set 11.5	DS11_05.dat	group [1=no reward, 2=delay reward;3=reward]; errors	Latent learning at day 17
ASCII file	DS11_06.dat	time1; time2	Stability of extroversion
ASCII file	DS11_07.dat	model [1=real life, 2=filmed, 3= none]; aggressn	Modeling aggression
ASCII file	DS11_08.dat	group [1=amnesic, 2=normal]; recall	Evaluation of amnesia
ASCII file	DS11_09.dat	oldwords; newwords	Implicit memory

Textbook designation	ASCII filename	Suggested variable names	Description
ASCII file	DS11_10.dat	group [1=amnesic, 2=normal]; score	Priming
ASCII file	DS11_11.dat	age; height	Age and height
Data Set 11.12	DS11_12.dat	pair; female; male	Matching attractiveness

Data Sets Used in Study Guide

Data Set	Suggested variable names	Description
CHILD_1.dat	familty [1=familiar, 2=stranger]; latency	Familiarity and compliance
CHILD_2.dat	perspive [1=condition 1, 2=condition 2, 3=condition 3]; rating	Perspective taking in parents
CHILD_3.dat.	authism; grade	Authoritarianism
CHILD_4.dat	chortime; concern	Household chores and children's concern for others
CHILD_5.dat.	mood [1=neutral, 2=good]; time	Mood and compliance
CHILD_6.dat.	attention [1=active, 2=nonactive]; complnce	Attention and compliance
CHILD_7.dat	perspive [1=condition 1, 2=condition 2, 3=condition 3]; parent; rating	Perspective taking in parents
CHILD_8.dat.	chortime; concern	Family ad hoc chores and children's concern for others
MEM_01.dat	T03,T06;T09;T12;T15;T18	Short-term forgetting
MEM_02.dat	conditn [1=read, 2=generate]; recall	Generation effect
MEM_03.dat	exptise [1=expert, 2=novice]; recall	Memory for chess positions
MEM_04.dat	exptise [1=expert, 2=novice]; recall	Memory for random chess positions
MEM_05.dat.	exptise [1=child expert, 2=novice adult]; recall	Chess expertise and children's memory
MEM_06.dat.	conditn [1=generate, 2=read-pair, 3=read]; recogn	Further test of generation effect
MEM_07.dat	conditn [1=generate, 2=read-pair, 3=read]; identif	Generation effect with different response measure
MEM_08.dat	level [1=appearance, 2= sound, 3= meaning]; recall	Levels of processing
MEM_09.dat	level [1=appearance, 2= sound, 3= meaning]; identif	Levels of processing in implicit memory
MEM_10.dat	level [1=appearance, 2= sound, 3= meaning]; instruct [1=incidental, 2=intentional]; recall	Levels of processing with incidental instructions
MEM_11.dat	conditn [1=read, 2=generate; partic]; recall	Test generation effect using a different design
MEM_12.dat.	conditn [1=generate, 2=read-pair, 3=read]; partic; recogn	Extended test of generation effect using a different design
MEM_13.dat	level [1=appearance, 2= sound, 3= meaning]; partic; recogn	Levels of processing using a different design

Data Set	Suggested variable names	Description
TEST_1.dat	aptscore	Developing an aptitude test
TEST_2.dat	extrovn	Extroversion scale
TEST_3.dat	school [1=school 1, 2=school 2, 3=school 3, 4=school 4]; aptscore	Possible differences in scholastic aptitude
TEST_4.dat	school [1=school 1, 2=school 2, 3=school 3, 4=school 4]; sex [1=female, 2=male]; aptscore	Possible sex differences in scholastic aptitude
TEST_5.dat.	apt_1; apt_2	Test reliability
TEST_6.dat.	aptscore; grade	Test validity
TEST_7.dat.	extrovn; aggressn	Extroversion and aggression